Quantum Themes
The Charms of the Microworld

Quantum Themes

The Charms of the Microworld

Thanu Padmanabhan

IUCAA, Pune, India

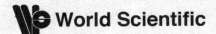 **World Scientific**

NEW JERSEY · LONDON · SINGAPORE · BEIJING · SHANGHAI · HONG KONG · TAIPEI · CHENNAI

Published by

World Scientific Publishing Co. Pte. Ltd.

5 Toh Tuck Link, Singapore 596224

USA office: 27 Warren Street, Suite 401-402, Hackensack, NJ 07601

UK office: 57 Shelton Street, Covent Garden, London WC2H 9HE

British Library Cataloguing-in-Publication Data
A catalogue record for this book is available from the British Library.

QUANTUM THEMES
The Charms of the Microworld

Copyright © 2009 by World Scientific Publishing Co. Pte. Ltd.

ISBN-13 978-981-283-545-1
ISBN-10 981-283-545-8

Printed in Singapore.

To the happy memories of the Science Society, Trivandrum

of the seventies

Preface

Quantum theory is one of the more abstract branches of theoretical physics; yet it makes clear and concrete predictions which are repeatedly verified in the lab. This fact, as well as its deep philosophical implications, has always fascinated the lay public, popular science writers and — of course — the researchers. More recently, there has been a significant amount of interplay and synergy between the concepts of microphysics and those of macrophysics. Currently fashionable ideas in cosmology — related to dark energy, structure formation, physics of the very early universe and even the creation of the universe — are deeply linked to concepts in quantum theory. This connection has given an additional impetus to this subject. Because of this strong background of interest, it is worthwhile to take stock of what quantum theory has achieved and where it is leading to at present, in a manner understandable to an educated layman.

Of course, there have been several such popular accounts written in recent years. I plan to make this book different from many others which are currently available, in the following aspects. To begin with, I have kept this book interesting but *not glamorous or mystifying*. There is an unfortunate tendency in the popular literature to do this while discussing the quantum theory. This appeals more to the sense of awe and wonder in the mind of the general reader; but, in the ultimate analysis, does not make the reader any more informed in the subject matter. The discussion is kept down-to-earth in the sense that the concepts will be described in a strictly scientific manner without trying to make them sound overly philosophical or "hyping" them up.

Second, I have kept the book reasonably up-to-date and focused on a key underlying theme. In order to do that I have decided to concentrate on the specific aspects of quantum theory which interface with gravity and cosmology. Hence, I have not included a variety of other topics (like, for example, the philosophical foundations of quantum mechanics, experimental details, technological applications ....) giving the book a clear focus and — of course — keeping it to reasonable length.

Finally, I have discussed all the aspects of science without diluting or distorting them. This would require careful explanation of the concepts and details (including the technical terminology) in a non-mathematical language. Very often authors tend to either talk down to the audience (saying it is too complicated to be explained in non-mathematical terms) or distorting the concepts in order to give the feeling of understanding to the average reader. I believe both these extremes need to be avoided and that a general reader should be treated with the respect he deserves.

This book is divided into 7 chapters. The first chapter outlines the concepts, successes and some history of classical physics from olden days till the end of the 19th century. The emphasis is on concepts and terminology rather than on the history, and the latter is introduced only to the extent that it illustrates some important ideas. The reader will be able to appreciate why the deterministic model of the universe was well-entrenched by the end of the 19th century and why several leading physicists felt that they only need to calculate more decimal places to understand everything about the universe. This chapter will form the backdrop for everything that follows.

In the second chapter, we start with the two "clouds" which were present in the horizon of classical physics, in the early years of the twentieth century and which led to the storms of quantum theory and relativity. Chapters 2 and 3 deal with them individually. In Chapter 2, I describe both special relativity *and* general relativity in a non-technical language and illustrate how they changed the philosophical foundations of classical physics. The emphasis, in the discussion of special relativity, is on those aspects which are required in the description of quantum field theory and particle physics later on. The emphasis in describing general relativity is twofold. First, it clearly points out how gravity acts as an "odd-man-out" and how several difficulties related to combining gravity with other interactions have their roots in the odd features of gravity. Secondly, it develops the concepts of general relativity to an adequate level for discussing cosmology and the early universe in Chapter 5.

The third chapter discusses the difficulties faced by classical physics which led to the development of quantum theory. I move directly to the fundamental concepts of quantum theory and explain how they are different from the corresponding concepts in classical physics. The discussion will be confined to the so called non-relativistic quantum mechanics at this stage since the ideas can be introduced most easily in that context.

The fourth chapter moves on to the issue of combining the principles of quantum theory and relativity, each of which has been described separately in the last two chapters. This leads one, in a natural fashion, to quantum field theory and particle physics. The concepts from particle physics, the success story of QED — the jewel in the crown — and the further developments in taming weak and strong interactions, all find their place here. I have also described topics like the Casimir effect which play a key role in the last few chapters.

Chapter five describes the connection between microphysics and macrophysics by exploring the role played by quantum effects in several epochs in the evolution of the universe. The thermal history of the universe — especially the nucleosynthesis and recombination — involves direct application of known quantum physics to understand key features about the universe. The inflationary phase, the choice for the dark matter candidate, the origin of initial fluctuations and the ever mysterious dark energy — all have quantum themes flowing through them and are discussed in this chapter.

One could say that the real conflict zone of today's theoretical physics is in the interface between quantum theory and gravity. This is taken up in Chapter 6, concentrating on several aspects of quantum gravity and quantum field theory in curved spacetime. The topics here include several issues (like the origin of black hole entropy) which are still unresolved but can be discussed in a fairly detailed manner based on the formalism developed in the earlier chapters. In a way, I will consider this chapter as the high peak of the climb through the book.

The short last chapter is distinctly more speculative than the rest of the book. But it addresses a question which has fascinated researchers and lay persons alike: How did it all begin? I do not pretend that we have all the answers but at the same time I do not want to downplay the significant advances we have made in understanding the laws of nature which allows us to ask questions which were once considered in the domain of philosophy and develop a framework in which they could be answered. I think this is very important and should be conveyed to the reader.

I have not included references to other works or to other popular books. This is a conscious decision since the increased access to the World Wide Web for the laypublic through powerful search engines makes such an exercise totally irrelevant. Anyone interested in these topics will be able to obtain the necessary information very quickly from the Web without me providing a list.

I thank Vasanthi Padmanabhan for dedicated support and for doing the texing and formating of the book as well as for producing most of the figures. I thank Sunu Engineer, Hamsa Padmanabhan, Sujata and Vinay Vaidya for reading earlier drafts of the manuscript and offering comments. I thank Gaurang Mahajan for help in generating one of the figures in Chapter 2. It was a pleasure working with the World Scientific Editorial team on this project.

T. Padmanabhan

Contents

xi

Chapter 1

The hubris of classical physics

1.1 Explaining nature: Apples and the moon

This book is about quantum physics which deals with, by and large, phenomena at very small scales. The way microscopic bodies like atoms, electrons etc. behave is quite different from what we are used to in our everyday life, based on the behaviour of tennis balls or motor cars. It came as a great surprise to physicists in the early part of the twentieth century that the microscopic world behaves in such a spooky way, so to speak, and it took some time for them to digest and accept it. Given this complexity, it is important to first understand how physical laws operate in the more familiar setting before plunging ourselves into the strange world of the quantum. This will be the aim of the present chapter.

It is probably best to approach this task through a series of simple examples. Let us begin by throwing a ball vertically up and catching it on its return. The ball starts moving upwards with some initial speed (which we will denote by v) and, as it goes up, its speed keeps decreasing. After some time, it reaches a maximum height (when its speed is instantaneously zero) and then it starts its downward journey. Its speed now increases and eventually it comes back to you with (roughly) the same speed as you threw it upwards. We are all familiar with this phenomenon. *The role of the physicist is to explain it.*

What does one mean by "explain" ? Well, to begin with you should be able to *predict* beforehand how high the ball will rise, how long it will take for the ball to come back etc.; more generally, we want predict its height x at any given time t; that is, you need to predict the function $x(t)$. You can't do any of these unless you have a 'rule'. (One usually says 'laws of physics' but ultimately they are just rules for calculating numbers which can be verified against observations). In this particular case, this is done using the rules of mechanics, which we will now describe.

The first question is what makes a body — which, in this case, is a ball — move. It turns out that you need a force to act on the body if its speed or direction of motion has to change. To simplify the description it is convenient to introduce at this stage the notion of *velocity*, which is essentially speed in a particular direction. The velocity can change when the speed of the body changes; but it can also change when the speed remains the same but its direction of motion changes. If you bounce a good rubber ball on the ground, its speed just before and just after hitting the ground will be the same but the direction of motion will change, thereby changing the velocity. Our rule says that you need to apply a force to change the velocity of a body; more importantly, it says you do *not* need any force to keep a body moving with a uniform velocity. This result is actually quite a deep one and has the names of Galileo and Newton associated with it. In contrast, Aristotle thought (wrongly) that you need a force to act on a body even to keep it moving with uniform velocity.

The rate at which the velocity changes is called *acceleration* and this is what you provide when you step on the gas in your car. The so called Newton's second law, which is the cornerstone of mechanics, says that the acceleration induced on a body by a force is proportional to the force. To state this more precisely, we introduce a quantity called the *momentum* for a moving body which — according to Newton — is given by the product of the mass of the body and its velocity; $p = mv$ where m is the mass of the body and v is its velocity. The rate at which momentum changes will be equal to the force that is applied. Taking the mass of the body to be a constant, this tells you that the force changes the velocity of the body; that is, force causes acceleration. So if you want to change the velocity quickly, you need to apply more force. Note that it is the rate at which the velocity changes which is important, not the velocity itself. You need to apply the same force to change the velocity of a body from 100 m s^{-1} to 101 m s^{-1} in 0.1 sec as to change the velocity from 1000 m s^{-1} to 1001 m s^{-1} in 0.1 sec.

Coming back to the example of the ball which you have thrown upwards, it is clear that you gave it an initial velocity. If there was no force acting on it, this velocity cannot change and the ball will continue to move upwards forever with the same velocity. However, we know that Earth is exerting a gravitational force on this ball, which changes its velocity. In this case, the force is acting downwards — in the direction opposite to the direction of initial velocity — and hence the force will decease the velocity. (This is called *deceleration* as opposed to acceleration). Eventually it will reduce the velocity to zero which is the instant at which the ball has reached the maximum height. From the next instant onwards, the downward pull on the ball gives it a velocity in the downward direction. Now the ball will start accelerating towards Earth and eventually you will catch it.

Given the acceleration induced on the ball by Earth's gravitational field — which happens to be about $g = 9.8$ m s^{-2} — you can compute how long it will take for the body to reach its maximum height. It is given by $t = v/g$; if you throw the ball with 9.8 m s^{-1} speed, it will take about 1 sec to reach its peak. We have done our first computation using a law of physics and predicted a result; someone can now check it with a ball and stop watch and either verify or disprove our law.

This example tells you the role played by physics in "explaining" nature. Here we start with the simple observation of how a ball behaves when it is thrown up. We then introduce certain laws: one to connect its motion with the force that is acting on it; and another to describe the force itself which — in this case — is the gravity of Earth. Given these two, we can explain the way any ball thrown upwards will behave! For our purpose, we only needed to say gravity produces an acceleration of 9.8 m s^{-2} on the ball but this is, of course, a very approximate statement. In different contexts, we need to use different laws.

The same analysis, of course, can be used to determine the behaviour of *falling* bodies. Suppose you drop a ball towards the ground from a tall tower of height h (Or suppose an apple falls from a tree of height h). How long will it take for the apple or ball to reach the ground and how fast will it be moving when it hits the ground ? Since the downward acceleration produced on the apple by Earth's gravity is g, it would have acquired a speed $v = gt$ in a time interval t. (Remember that acceleration is just the rate at which speed changes so that $g = v/t$). If the apple took an amount of time T to hit the ground, its final speed will be $V = gT$ while initially it had zero speed. So the average speed with which it moved is $(1/2)(gT + 0) = (1/2)gT$. With this average speed, it would have covered a total distance $(1/2)gT \times T = (1/2)gT^2$ which we know is the height h. So we get the relation $h = (1/2)gT^2$. If we know h and g, this relation tells you how to determine the time it takes for an apple to fall to the ground.

A more nontrivial example is given by the motion of Moon around Earth. The Earth exerts a gravitational force on the Moon making it fall towards Earth, just as it does to an apple. The notion of Moon 'falling'

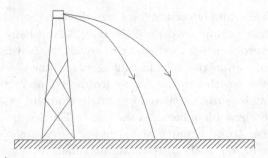

Fig. 1.1   The trajectory of a ball thrown horizontally from the top of a tower. It will fall at some distance away from the foot of the tower. If the initial speed is increased it will fall farther away. In this figure we have assumed that Earth is flat.

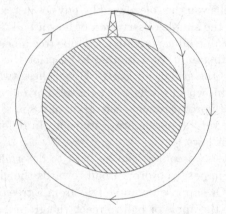

Fig. 1.2   The trajectory of a ball thrown horizontally from the top of a tower taking into account the curvature of Earth. (The dimensions are quite exaggerated.) As the initial horizontal speed of the ball is increased, the ball travels a larger distance before hitting the ground. At a critical value of the speed, the ball misses the ground completely because the Earth 'curves away'. The ball now will be in a circular orbit around the Earth. This shows that an artificial satellite (or moon) orbiting the Earth is actually falling towards the Earth all the time in its orbit.

towards Earth may be a bit strange; so we will explain that idea first. To see what is involved, consider a stone thrown *horizontally* from the top of a tower (see Fig. 1.1). We know from experience that it will fall at some distance away from the foot of the tower. If you throw it with larger speed, it will fall farther away from the foot of the tower. Figure 1.2 shows the same situation but now taking into account the fact that Earth is not flat.

It is now clear that, as we increase the speed of the stone, a limit will be reached when the stone simply goes round and round the Earth! Of course, the stone is all the time trying to fall to the ground but the Earth is 'curving away' preventing this from happening. This is precisely how you place a satellite in an orbit. Instead of throwing it from a tower, we take the satellite up in a rocket and then give it the appropriate horizontal speed so that it will keep falling to Earth without ever hitting it. The Moon, of course, is just a natural satellite orbiting the Earth in the same manner. It is constantly falling to the Earth, just like an apple from the tree.

Earth induces an acceleration of 9.8 m s^{-2} on the apple but a much smaller acceleration on the Moon. It turns out that, according to Newton's law of gravity, the acceleration induced by a body of mass M on another body at a distance r is $g = GM/r^2$ where G is a constant. So the acceleration increases in proportion to the mass but decreases as a square of the distance. That is, when the distance is doubled, the force falls by a factor $2 \times 2 = 4$; if the distance is tripled, it falls by a factor $3 \times 3 = 9$ etc. Such a force law is said to satisfy the "inverse square" relation. Since the Moon is nearly 60 times farther away from the centre of the Earth than the falling apple, the acceleration is only about 0.0027 m s^{-2} (see Fig. 1.3). You can now use this rule to calculate how exactly the Moon should move around Earth and — of course — the result matches with the observations.

You should pause for a moment to think through the miracle we have achieved. Given the two laws (one relating motion to force and another giving how the acceleration due to gravity varies with distance) we have unified a purely terrestrial phenomenon (falling of apple or a ball) to an awesome celestial phenomenon (motion of Moon around the Earth). *This is the true spirit of physics:* It provides verifiable links between apparently unrelated phenomena and allows one to describe a large number of facts using a handful of 'rules' and, of course, a lot of maths which we have skipped! We no longer need two explanations for the apple falling towards Earth and Moon going around the Earth — they are fundamentally the same.

While the real drama of nature in the above examples happens through changes (changing position of bodies, changing velocities .....) one of the fundamental considerations in physics is to sift through this and determine physical quantities that remain constant. In the cases we considered above it turns out that there is one physically important quantity, viz. the total energy, which is conserved — that is, it does not change during the motion. Take the case of the ball that is thrown up which has two different kinds

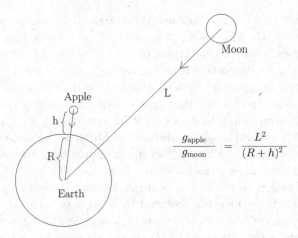

Fig. 1.3   Earth's gravitational force affects both the apple falling toward the Earth and the Moon which is falling toward the Earth at each instant in its orbit (see the previous figures). Given the distance to the Moon, one can compute the gravitational forces and relate the motion of the Moon to the motion of the apple.

of energies. The first is the kinetic energy that arises due to the motion and is given by the formula $K = (1/2)mv^2$ where m is its mass and v is its speed. Since the speed of the ball keeps changing, this kinetic energy is not a constant. The second one is called the potential energy which is due to the location of the ball with respect to the ground and is given by $V = mgx$ where x is the height above the ground (which is taken to be positive) and g is the acceleration due to gravity (see Fig. 1.4). Initially, $x = 0$ making the potential energy zero while the ball has some kinetic energy due to its initial speed. As the ball rises to its maximum height, its speed and kinetic energy decrease; but its potential energy increases because x increases. The miracle is that the sum of kinetic and potential energies, which is the total energy, remains constant during the motion. When the ball is in its downward trip, its kinetic energy keeps increasing while its potential energy keeps decreasing keeping the total energy again constant. We will see in later chapters that a good part of physics is a quest for such conserved quantities. (Incidentally, the expressions given above for kinetic and potential energies are approximate — as we will see in the next chapter — but the fact that the total energy is conserved is an exact statement.)

Energy is just one of the many quantities which are conserved during

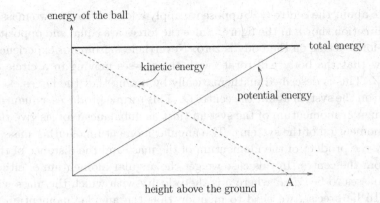

Fig. 1.4 Conservation of energy in the motion of a ball thrown upwards from the ground. As it rises up its kinetic energy decreases and potential energy increases keeping the sum constant. On reaching the maximum height at A, the kinetic energy is zero and the total energy is equal to the potential energy. On its way down, the potential energy decreases with the kinetic energy increasing.

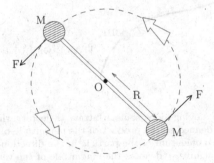

Fig. 1.5 The concept of angular momentum for a simple rotating system. When a force F is applied on each of the two masses connected by a rod as shown in the figure, the system will rotate about the centre O. If each of the masses move with a speed v, the angular momentum of either mass is defined to be MvR.

the time evolution of a system. Another quantity which is often conserved — and will play a crucial role in our description of quantum mechanics — is called *angular momentum*. Angular momentum of a rotating body, in Newtonian mechanics, is essentially the sum of the momenta of the particles in the body but weighted by the distance of the particles from the axis of rotation. Consider, for example, the dumbell shaped body in Fig. 1.5 made of two masses M connected by a rigid rod of length $2R$. The system is free

to rotate about the centre O. Suppose we apply a force F to the two masses
in the direction shown in the figure. Since the forces are equal and opposite,
the net force acting on the body is zero. Nevertheless, from our experience,
we know that the body will rotate with the masses moving in a circle of
radius R. This is described mathematically by noting that the forces exert
a torque on the system about the centre O. This torque produces a quantity
called angular momentum of the system (just as unbalanced forces give rise
to the momentum of the system). The angular momentum of either mass is
given by the product of the momentum of the mass and the distance of the
mass from the centre. In this case we get the angular momentum of either
of the masses to be MvR where v is the velocity with which the mass will
move. To be precise, we need to mention that the angular momentum is
also a vector and — in this case — it is directed along the direction of the
axis of rotation.

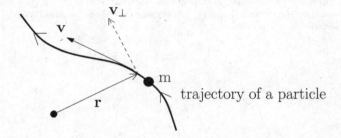

Fig. 1.6　For a particle moving along some arbitrary trajectory, the angular momentum
at any given instant is defined as the product of the magnitude of the position vector $\mathbf{r}$
of the particle and the momentum of the particle in the direction perpendicular to the
position vector. In the figure $v_\perp$ denotes the magnitude of the velocity in the direction
perpendicular to the position vector and hence the angular momentum is $mv_\perp r$.

As a second example consider a ring of mass M and radius R, rotating
uniformly about an axis passing through the centre and normal to the plane
of the ring. If each particle in the ring is moving with speed v we attribute
to the rotating ring the angular momentum MvR. You see that one can
increase the angular momentum by increasing any of the three quantities
M, v or R. In the case of a more complicated motion like the one shown
in Fig. 1.6, the angular momentum of the particle at any given instant
is calculated by the product of the mass, the distance from some chosen
origin and the magnitude of the velocity in the direction perpendicular to
its position vector. In Fig. 1.6 the quantity $\mathbf{r}$ denotes the position vector

of the particle, **v** denotes the velocity and the dashed arrow indicates the magnitude $v_\perp$ of the velocity in the direction perpendicular to **r**. For an isolated system, the angular momentum is a conserved quantity, just like energy or momentum.

The energy and angular momentum are just two examples of quantities which are conserved in physical systems. The term "conservation" in physics means that a physical quantity retains its numerical value when the system evolves in time. In the example we saw above, a stone moves upwards from Earth, reaches a maximum height and falls back to Earth causing its position, speed, kinetic energy etc. to change from instant to instant. But the total energy of the stone does not change during its motion. In other areas of physics, one comes across much more complicated conservation laws leading to other physical quantities (like energy and angular momentum in mechanics) which remain conserved during physical processes. These conservation laws and their discovery are very important in the development of physical theories. Quite often one can characterize the properties of a theory by the quantities which remain conserved in the theory. We shall see several examples of more complicated conservation laws in the later chapters.

1.2 More laws of classical physics

Mechanics, which essentially deals with the motion of terrestrial and celestial bodies, was one of the earliest areas of natural phenomena in which physics made a mark. However, it was known right from antiquity that there are other phenomena in Nature which are not purely mechanical. These observations have also played a crucial role in the development of physics and we will now take a quick look at some of them. Electric and magnetic phenomena probably deserves the centrestage in such a discussion.

Electrical forces, which act between particles carrying electric charge, were known to mankind for a very long time. You probably will recall from your high school days that a glass rod, when rubbed with a cloth will attract small pieces of paper — which is an example of the electric force in action. A body can have positive, negative or zero electric charge. (The last category is called "neutral".) A positively charged particle will repel another positively charged particle while attracting a negatively charged particle; similarly a negatively charged particle will repel a negatively charged particle but will attract a positively charged particle. Neutral particles do not feel any electrical force. The force of attraction or repulsion increases in direct proportion with the electric charge of the particle. That is, a particle carrying twice the amount of electric charge will exert a force which is twice as large. The force decreases with increasing distance between the charged particles as the square of the distance between them, just as in the case of gravity.

Like the electric forces, magnetic forces were also known to humanity for centuries. For example, many ancient civilizations knew that a compass needle will stay north-south because of its magnetic properties. For several centuries these electric and magnetic phenomena were treated as separate entities: Certain phenomena required electrical forces for their explanation and others needed magnetic forces. What was not known in the early days was that electric and magnetic forces are essentially the same. It turns out that magnetic force is produced by moving electric charges and is not a separate kind of force caused by some new kind of "magnetic charge". The moving charges constitute what is usually called an electric current and it is this electric current which produces the magnetic field. Thus one should view the electromagnetic force as a single entity.

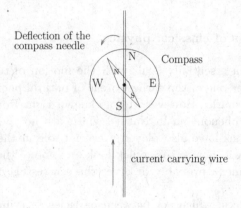

Fig. 1.7   Electric current can produce a magnetic field, as discovered by Ampere. The current which flows through the wire produces a magnetic field that acts on the compass needle and deflects it.

The first two breakthroughs indicating this inter-relationship between electric and magnetic forces came when it was realized (by the scientists Ampere and Faraday, among many others) that electricity can produce magnetism and vice versa. For example, if you hold a straight piece of wire above a compass needle (which responds to magnetic forces) and send a current through the wire, the compass needle will be deflected. This is a clear proof that an electric current creates a magnetic field around it (see Fig. 1.7). On the other hand, if a magnet is moved in and out of a coil of wire, it generates an electric current which will flow through the coil — which lies behind most procedures to generate commercial electricity (see Fig. 1.8).

These two features, called Ampere's law and Faraday's law, certainly show that electric and magnetic phenomena are closely related and are — in a precise, specific sense — interconvertible. There is, however, one crucial difference. Faraday's law says that a *changing* magnetic field produces an electric field. On the other hand, Ampere's law connects a *steady* current to a magnetic field without the existence of any changing electric field. To see this difference more clearly, note that there are two sources for the electric field: one is just the charged particles (as in the case of a glass rod rubbed by a cloth) and the second is the changing magnetic field, discovered by Faraday. If there has to be some symmetry between electric and magnetic phenomena, we should expect two possible sources for the magnetic field as well. The first is a steady current, which is what Ampere found. The second, which we are now contemplating based purely on symmetry considerations, should be a changing electric field. That is, we would expect a *changing* electric field to produce a magnetic field just as a changing magnetic field leads to an electric field. The first part of the story — electric current producing the magnetic field — was already in place but what was not clear was whether a changing electric field can also produce a magnetic field.

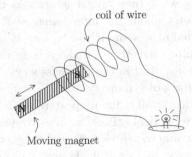

coil of wire

Moving magnet

Fig. 1.8   Faraday's law: A magnet moving through a coil can produce an electric current flowing through the coil.

It took the genius of J.C.Maxwell, the British physicist, to assert that it does. Somewhat indirectly, Maxwell was looking for symmetry in the mathematical equations describing the electric and magnetic phenomenon. The set of laws, as available at that time, had one major problem: They were not consistent with the law for conservation of charge. Roughly speaking the conservation of charge means the following: Suppose you find that there is a certain amount of charge in a particular region of space at a given instant of time. If the charges move in or out of this region, you keep tabs on the amount of charge transfer. Since electric current is nothing but moving charge, this is easily done by calculating the electric current flowing through the surface enclosing this region of space. When you count the total amount of charge in the region at a later time, it should be precisely accounted for by the initial amount of charge plus the amount which flowed into the region minus the amount which flowed out of the region. In other words, you cannot create or destroy charge but can only make it move from place to place in the form of an electric current. Since the currents and charges are intimately linked to electric and magnetic fields, you can imagine that any law obeyed by the latter will have something to say about charge conservation. It turns out that to maintain the conservation of charge (which is a fact established by observations) we need to change Ampere's law and introduce the new feature: viz., that a changing electric field can lead to a magnetic field. This is precisely what Maxwell did. The symmetry between electric and magnetic fields was now complete.

Figure 1.9 gives the laws governing the electric and magnetic fields in empty space in modern notation. The symbol **E** stands for the electric field and the symbol **B** stands for the magnetic field. The other funny symbols in the figure refer to mathematical operations which you are free to ignore! Even without knowing what they are, it is obvious that these equations have a sense of symmetry about the interchange of **E** and **B**. Except for a minus sign, they remain the same if you replace **E** by **B** and vice versa. A physicist will — quite correctly — make a song and dance about the beauty in these equations. And in the physicist's eye, beauty is invariably associated with some form of symmetry.

You might think this is all a bit preposterous and wonder why what Maxwell did was such a big deal. "Fine, he made the equations which have weird symbols more symmetric in electric and magnetic fields, so what?" you might ask. It *is* a very big deal; if Maxwell's new term was not there much of the technology we take for granted in the modern world will disappear! The new term led to the existence of electromagnetic waves —

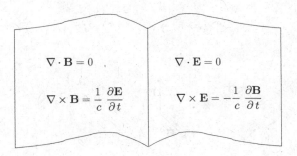

$$\nabla \cdot \mathbf{B} = 0 \quad , \qquad \nabla \cdot \mathbf{E} = 0$$

$$\nabla \times \mathbf{B} = \frac{1}{c}\frac{\partial \mathbf{E}}{\partial t} \qquad \nabla \times \mathbf{E} = -\frac{1}{c}\frac{\partial \mathbf{B}}{\partial t}$$

Fig. 1.9   The Maxwell's equations written in modern notation. These equations govern the behaviour of source-free electromagnetic field and show the intimate relationship between electric and magnetic phenomena. Even without knowing what the symbols mean, you can see that these equations remain the same, except for a minus sign, if you replace $\mathbf{E}$ by $\mathbf{B}$ and vice versa.

something we use in every form of communication today, to say the least.

Let us try to see how this come about. Suppose you set up an electric field in some region of space, possibly using currents and charges. Now suppose the electric field tends to die down when you switch off the sources. Maxwell's equation — because of the crucial new term he added — will now predict the generation of a magnetic field because of the *changing* electric field. So when the electric field dies down it would have generated a magnetic field; now if the magnetic field dies down, it will generate an electric field due to Faraday's law! So we now open up the possibility of the existence of oscillating electric and magnetic fields, each generating the other and surviving to all eternity! What you now have is an *electromagnetic wave* (see Fig. 1.10) with a life of its own. R.P. Feynman describes this phenomenon as the caterpillar becoming a butterfly!

There is more to the genius of Maxwell. When you work out the maths, it turns out that such an electromagnetic wave will move with a speed that can be calculated from Maxwell's equations and known physical constants. When this is done, one finds that the speed of the electromagnetic wave is precisely equal to the speed with which ordinary light was (known to be) travelling! At that time no one had a clue what light was — and, least of all, no one suspected that light has something to do with electricity or magnetism. So this was a great moment for physics. Not only did Maxwell unify electricity and magnetism into a single structure but — in the process — he explained light as an electromagnetic wave.

Since we will be dealing extensively with the wave phenomena later on,

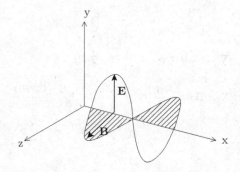

Fig. 1.10   The electromagnetic wave is made of oscillating electric and magnetic fields orthogonal to each other. The wave propagates in a direction perpendicular to both the fields.

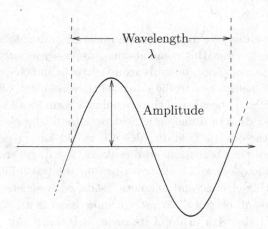

Fig. 1.11   Basic parameters which describe a wave.

this is a good place to get acquainted with some elementary properties of waves. One can usually associate with any wave four different physical quantities which are illustrated in Fig. 1.11. (For the sake of definiteness, let us consider an electromagnetic wave which is essentially an oscillating electric or magnetic field in space; but most of our discussion is applicable to any wave.) The first quantity is the wavelength which is the distance over which the wave property repeats itself; it can be measured as the distance between two consecutive crests (or two consecutive troughs) and is usually denoted by the Greek letter λ ('lambda'). The next is the frequency of the wave which measures the number of oscillations per second taking place at any given location and we will denote it by the symbol ν ('nu'). The third quantity is the speed with which the wave is propagating forward; in case of electromagnetic radiation, this speed is the same as speed of light, c, — remember that light is a form of electromagnetic radiation. The last physical attribute for a wave is called amplitude which describes how high the crests of the wave rise or the troughs deepen (see Fig. 1.11).

From the definitions, it is easy to see that the first three quantities are related by $c = \nu\lambda$. Since c is a constant, the frequency of electromagnetic radiation will increase when its wavelength decreases and vice versa. The nature of the electromagnetic wave and its physical attributes depend sensitively on its wavelength so that it is often difficult to imagine that they are all the manifestation of the same underlying phenomenon. Among the well known types of electromagnetic radiation the ones with largest wavelength are radio waves without which modern communication will come to a standstill; next comes the microwaves which are produced, for example, in your kitchen oven; then we have infrared radiation by which heat is transmitted to you from your fireplace; next is the normal visible light which our eyes are sensitive to with the wavelength decreasing from red light to blue light; the ultra violet radiation is that which tans your skin on a summer holiday; at still higher frequencies we have the x-rays which your dentist uses and finally, the radiation with the tiniest of the wavelengths called gamma rays which are fairly harmful to our existence. This short summary should caution you that the same physical entity — an electromagnetic wave — can have completely different properties depending on its wavelength or, equivalently, its frequency.

We described above how an electromagnetic wave — once created — has a life of its own and is self-sustaining. The decay of either the electric or magnetic field leads to the other. But that did not answer the question of how we generate an electromagnetic wave in the first place. It turns out that, at the basic level, any charged particle that is accelerated will emit electromagnetic radiation. So if you push a charged particle, making it move with some acceleration, you would have generated an electromagnetic wave. If you now stop the charge from accelerating, it will stop generating electromagnetic radiation but what has already been generated will keep propagating outwards at the speed of light. In more practical contexts (like in a radio transmitter which produces electromagnetic radiation) , the acceleration of charges will be more complicated and precisely coordinated but the basic principle is the same.

This is also probably a good place to mention the serious trouble this result causes in describing microscopic, atomic systems. Early experiments suggested that the best description of an atom is as a system with a positively charged nucleus at the centre and electrons (which carry negative charge) orbiting around it. An electron orbiting in a circular path, say, is of course accelerating. (Recall that, in a circular motion the *direction* of the velocity is continuously changing due to the acceleration produced by the force of attraction between the electron and the positively charged nucleus at the centre.) Our theory says that such an electron should be radiating electromagnetic waves and thus will be losing energy. As this happens, the radius of the electron orbit will shrink and the electron will spiral down to the nucleus. Calculations show that this will happen in an extraordinarily small time interval. So, according to classical electromagnetic theory the atoms will collapse in no time and no structure can exist! We will see later in Chapter 3 that this is one of the reasons we need the classical theory to be modified by quantum mechanical considerations.

1.3   Laws of physics in moving cars

Now that we have introduced the laws of mechanics and laws of electrodynamics, it is time to take a look at some of the broader issues related to the nature of physical laws. Let us begin with mechanics. The key feature in Newton's law of motion is that motion with a uniform velocity does not require any force. This innocuous looking statement has some deep implications which we will now highlight.

Consider a car moving on a straight road with constant speed. Suppose you are inside the car and drop a coin vertically down. If the car was not moving, the coin will, of course, fall vertically down on a spot A directly below the point at which it is released (see Fig. 1.12). The question is where it will fall in the moving car: a little behind A, little to the front of A or still on the spot A? At first you might think as follows: "The coin takes some amount of time to hit the floor of the car. During this time, the floor of the car would have moved forward a little bit; so the coin has to fall at a point which is little behind A." This, however, is incorrect and the coin will fall exactly on A, as though the car is not moving.

The reason is as follows: When you release the coin from your fingers, it has a velocity in the forward direction which is same as that of the car. After the release, Earth's gravity acts downwards and gives the coin a downward acceleration *g*. This will lead to a component of velocity in the downward direction which will slowly increase with time and the coin will eventually hit the ground. This vertical motion of the coin is the same in both a stationary car and in a moving car. But note that there is no force acting on the coin in the horizontal direction. This means that the horizontal speed of the coin cannot change; in particular it can't be reduced to zero from the initial value it had before you released it from your fingers. So the coin continues to move forward exactly at the same speed as the car, even after it is released! Obviously, the spot A, which is also moving forward because of the car's motion, will always be directly under the coin as the coin falls. Eventually the coin will land at A!

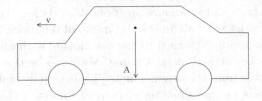

Fig. 1.12 A coin is dropped from inside a car such that it falls on the point A at the floor of the car when the car is not moving. If the experiment is repeated with the car moving with uniform velocity, you will find that the coin again falls at A. If the car was moving forward with an *acceleration*, the situation will change and the coin will fall behind A — that is a little to the right of A in the figure.

If this were not true, you could have determined whether the car is moving with uniform speed or not without looking out of the windows. You first mark the point A when the car is stationary. Then whenever you want to know whether the car is moving or not, you just drop a coin. If it falls behind, you conclude that the car is moving. Our analysis above shows that this procedure won't work. The coin will fall on A whether the car is moving with uniform speed or not. You cannot distinguish between a car at rest and a car moving with uniform velocity by dropping a coin. You may ask: "Fine, dropping coins won't be enough; but is there some other experiment I can perform inside the car to determine whether the car is at rest or moving with uniform velocity?" It turns out that, because forces care only about accelerations and not about velocities, you cannot devise any mechanical experiment inside a car to determine whether the car is moving with uniform velocity or not.

Note that all these comments are only applicable to motion with *uniform* velocity. If the car is *accelerating* forward, the coin will fall slightly behind A. Now the coin will still move forward with the speed it had when it was released (which is the same as the speed of the car at the instant when you released your fingers) but the car is accelerating forward. So the spot A will move forward with an acceleration while the coin will move forward with only the initial speed it had and no acceleration. Obviously, the spot A will be ahead of the coin all the time; in other words, the coin will now fall behind the spot A.

All these show that motion with uniform velocity is rather special. Physicists like to talk in terms of *frames of reference*, which is just an abstract way of thinking about moving cars. Suppose there is some region of space in which Newton's laws of motion hold and we introduce three coordinate axes (x, y, z) to measure positions of particles. This will constitute what a physicist will call an inertial frame of reference. Consider now a coordinate grid rigidly attached to the car moving with uniform velocity with respect to the original inertial frame. We just saw that such a frame is indistinguishable from the original one. This is usually stated as follows: *Laws of mechanics are the same in all inertial frames.* (This is called the Galilean principle of relativity.) Such a result constitutes what physicists call a symmetry principle or an invariance principle. Broadly speaking if your results do not change when you change some external circumstance, you have a symmetry principle in operation.

Laws of mechanics remain the same in all inertial frame but what about other laws? For example, does the laws of electrodynamics — encoded in Maxwell's equations — remain the same in all inertial frames? Well, the answer turns out to be complicated and — in fact — will form the cornerstone of relativity we will discuss in the next chapter. The laws of electrodynamics are indeed invariant when you go from one inertial frame to another; but it turns out that the *relation* between the two inertial frames needs to be defined in a new manner for this to work. If you use this new definition with laws of mechanics, then it turns out that laws of Newtonian mechanics are *not* the same in all inertial frames! In other words, you cannot define the idea of inertial frames mathematically such that *both* the laws of Newtonian mechanics and electrodynamics remain the same in all inertial frames.

Another way of stating this result is as follows: Let us assume that Newtonian mechanics is correct and that laws of mechanics are the same in all inertial frames. We will then find that laws of electrodynamics are not the same in the inertial frames, defined using the results for mechanics. In practical terms, this means that you can use electromagnetic phenomenon — like the propagation of light — to determine whether a car is at rest or moving uniformly. People tried to do this using the Earth itself treated as a car moving through space. All these experiments led to a null result! That is, one simply could not determine the motion of a frame using laws of electromagnetism even though theory said this should be feasible. It took the genius of Einstein to change the theory and with it the mathematical relation between the inertial frames. The result was the special theory of relativity which we will discuss in the next chapter.

1.4 Heat as a form of motion

There is another domain of great practical importance in which physics could make significant progress before the twentieth century. This has to do with the the nature of heat and thermal phenomena which, when we look closely, turn out to be quite nontrivial.

The simplest of these examples deals with the properties of bulk matter. If you take a solid body and heat it, its temperature will increase — something even a caveman knew after the invention of fire. The concept of a body being hot or cold and even the idea of putting heat to use (like in, say, cooking) is as old as fire and was known from prehistoric days. What is more, the body can eventually melt and become a liquid if you heat it enough. We are all familiar with a chunk of ice turning into water when heated. If you heat it further, the liquid can vaporise and become a gas. (In the case of water it becomes steam.) Though ice, water and steam have completely different physical properties, it is clear that — in a fundamental sense — they are all the same. For one thing, if you freeze the water obtained by melting ice, you get back ice; but not, say, silver. So there is some kind of a 'memory' running through these three states.

The real question is: what makes a hot body hot? To be concrete let us consider two identical blocks of, say, copper, one kept at 30° C and another at 100° C. Anyone can see that the physical effects of these two on the surrounding will be different (just touch them!) but what exactly is different in these two blocks? For a long time people thought that heat is some kind of 'fluid' and heating a metal increases the content of this fluid in the metal. If you dip a hot copper rod into a pan of water at room temperature, say, the metal will cool and the water will get heated; eventually both will reach the same temperature. This was thought to be due to the transfer of the 'heat fluid' from copper to water.

Very soon experiments showed that such a view leads to several difficulties. Most importantly, this 'heat fluid' needed to have all kinds of peculiar properties if it should be able to move invisibly from one body to another and remain in it without altering any of its properties. Further, in this approach, it is not very easy to explain several simple relations that were known about heated gases. For example, consider a gas kept in a container of volume V at temperature T. The pressure P exerted by this gas will vary inversely as the volume of the gas if the temperature is constant. That is if you halve the volume, the pressure will double, when temperature is kept constant. (This is known as Boyle's law). Similarly, if you double the temperature the pressure will double, if the volume is kept constant (which is known as Charles' law). One can combine these results to say that $PV \propto T$ for a gas. The real question is how do we understand these laws? Are these fundamental laws we need to add to our collection or can one understand these from something which we already know? Remember that this is the real spirit behind physics. Once we understood the falling apple using gravity we could also explain the motion of Moon around Earth — we did not need one law for the apple and one for the Moon. In the same way, can we understand the thermal behaviour of a gas from something which we already know?

Indeed we can and this correct interpretation of the nature of heat was arrived at by Boltzmann. (Unfortunately his views came under a lot of flak, eventually driving him to suicide; physics is a dangerous vocation at times.) Boltzmann suggested that heat is just random motion of microscopic constituents of a body. We know *today* that all matter is made of atoms or molecules but, of course, this was far too radical during Boltzmann's days. In a solid, one can imagine these atoms to be arranged in some kind of a lattice structure but these atoms are not stationary. They vibrate about their equilibrium position with a small amplitude. One can therefore associate some amount of random energy with these atoms in the solid lattice. The temperature of the solid is essentially a measure of the average random kinetic energy of the atoms; the larger their random kinetic energy is, the larger is the temperature. If you raise the temperature too much the random vibrations will increase to such an extent that the lattice structure will be destroyed and the solid will melt becoming a liquid. Of course, the atoms still remain in the liquid, with random motions and if you further heat the liquid, their random motion will increase. Eventually, the liquid will vaporise into a gaseous state (like water becoming steam). In the gaseous state, each atom (or molecule) will again have random motion which will continue to be a measure of its temperature.

As an aside, let us mention that the lowest energy state for any matter occurs at around −273 degree centigrade and it is impossible to cool any body to temperatures lower than this value. (To cool a body, we have to extract the heat energy from the body; which is impossible if it is already in the lowest energy state). In physics, it is often convenient to use a different unit so that the lowest temperature corresponds to zero degree. This unit is called "Kelvin" and is defined so that zero degree Kelvin (0 K) corresponds to −273 degree centigrade. To obtain the temperature in Kelvin, we merely have to add the number 273 to the temperature expressed in centigrade. Thus the temperature −183 degree centigrade (at which oxygen becomes a liquid) corresponds to $-183 + 273 = 90$ Kelvin.

Boltzmann's insight was to reduce the — until then mysterious — thermal phenomenon into one of mechanics. If a gas is made of zillions of randomly moving molecules and the temperature of the gas is linked to the random kinetic energy of motion, then the study of thermal properties of the gas reduces to the study of the random motion of the molecules. That is, thermodynamics now becomes part of Newtonian mechanics. Since we will be dealing with a huge number of molecules at any given time, the emphasis shifts from describing the trajectories of each molecule or atom to describing the statistical properties of the system. Hence this branch of physics, which originated from Boltzmann, is often called *statistical mechanics*. It tells you how to link up the bulk properties of matter — like, for example, the pressure of the gas — to the microscopic motion of the molecules.

Just to see how this works, let us consider pressure exerted by a gas of N molecules kept in a cubical box of size L, using nothing more than high school algebra. (If it scares you, skip ahead to next paragraph!). We will concentrate on the side of the box perpendicular to the x-axis. The gas exerts a pressure on the wall of the box because the collision of the molecules with the wall exerts a force on the latter. Let us assume that each molecule is moving randomly with an average speed v. In a time interval t, the molecules can travel a distance vt on the average; so all the molecules which are at a distance less than vt from a wall will collide with a wall in the time interval t (see Fig. 1.13). This number is given by the molecules which are inside a volume $(1/6) \times vt \times L^2$ where the factor $1/6$ takes into account the fact that only a third of the molecules will be moving along the x-axis and only half of these will be moving *towards* the

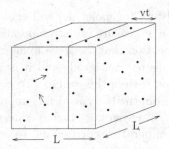

Fig. 1.13   One can compute the pressure exerted by a gas from Newtons laws of mechanics, if we assume that the gas is made of large number of molecules in random motion. This hypothesis, made by Boltzmann, interpreted heat as a form of motion and essentially unified the laws of thermodynamics with laws of mechanics.

wall. Since there are (N/L^3) molecules per unit volume, this number is $(N/L^3) \times (1/6)vtL^2 = Nvt/6L$. On colliding with the wall the molecules will bounce back so that the momentum of each one changes from mv to $-mv$; that is by an amount $2mv$. The total change of momentum for all the molecules is $(Nvt/6L) \times 2mv = (N/3L)mv^2t$. Since this occurs in the time interval t, the rate of change of momentum, which is the force exerted on the wall by the molecules, is obtained by dividing this expression by t giving $F = (N/3L)mv^2 = (2N/3L)K$ where $K = (1/2)mv^2$ is the kinetic energy of each molecule. The pressure of the gas is defined as the force it exerts per unit area of the surface; here the area that is involved is L^2 and hence the pressure is $P = F/L^2 = (2/3)K(N/V)$ where $V = L^3$ is the volume of the box. This 'law' can be written as $PV = N(2/3)K$ which relates the pressure of the gas, volume of the gas and the mean kinetic energy of the molecules. It was known, on the other hand, that for a gas with fixed number of particles, the product PV is actually proportional to its temperature. Boltzmann's insight was in identifying the notion of temperature with the mean kinetic energy of the particles and suggest $K = (3/2)k_BT$ where T is the temperature and k_B is called — quite appropriately — Boltzmann's constant. This direct relation between kinetic energy of molecules and the temperature allows us to write our earlier relation as $PV = Nk_BT$ which, some of you might recognize as the equation describing an ideal gas.

   This is again something truly remarkable and one needs to savor the moment. In thermodynamics, one would have described a gas in a container without ever worrying about whether it is made of molecules or not. The gas can be a continuous substance with no substructure and we can still define its pressure,volume etc. and study its properties. In this case, however, we will have no fundamental understanding of why the product PV for a gas is proportional to its temperature; we need to postulate it separately as a new law. Boltzmann's idea was that if we assume the gas is made of large number of particles and we use for these particles the usual laws of mechanics (recall that we just computed the rate of change of momentum due to collisions with the walls) then we can describe how such systems behave. In the process, we realize that temperature has a simple interpretation as the average kinetic energy of the molecules and we no longer need to introduce any mysterious 'heat fluid'. Just as Maxwell's unification allowed one to understand light as an electromagnetic phenomenon, Boltzmann's work allows us to understand thermal phenomena as a part of Newtonian mechanics.

Given all these, the current picture of bulk matter can be understood as follows. Consider, for the sake of definiteness, a solid piece of ice with which you are quite familiar in everyday life. Ice, like most other solids, has a certain rigidity of shape because, in a solid, atoms are arranged in a regular manner. Such a regular arrangement of atoms is called a "crystal lattice" and one may say that most solids have "crystalline" structure (see Fig. 1.14). Atoms, of course, are extremely tiny and they are packed fairly closely in a crystal lattice. Along one centimeter of a solid, there will be about one hundred million atoms in a row. The standard notation for this is to say that there are 10^8 atoms along one centimeter of ice. (The notation 10^8 stands for ten multiplied eight times, which is a number with 1 followed by 8 zeros.) This means that the typical spacing between atoms in a crystal lattice will be about one part in hundred millionth of a centimeter, i.e., about $1/100,000,000$ centimeter. This number is usually written as 10^{-8} cm. The symbol 10^{-8}, with a minus sign before 8, stands for one part in 10^8; i.e., one part in $100,000,000$. (In this notation — which we shall use frequently — one tenth will be 10^{-1}, one hundredth will be 10^{-2} etc. To denote two parts in a thousand, we will write 2×10^{-3}. Note that 3×10^3 stands for 3000 while 3×10^{-3} stands for $3/1000$, viz. three parts in a thousand.) To describe the phenomenon at atomic scales, we have to use such small lengths and it is useful to introduce new units to describe very small lengths. The length, 10^{-8} cm, is a convenient unit and is called one Angstrom (1Å, for short). So the separation between the atoms in a crystal is about one Angstrom. A small cube of a solid, 1 cm in the side, will contain about 10^{24} atoms. It is because of this tiny separation between the atoms that solids appear continuous and people were skeptical when

Boltzmann introduced the substructure to matter.

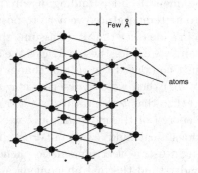

Fig. 1.14   Many solids are made of a regular array of atoms in the form of a lattice. The figure shows the 'crystalline structure' of such a simple solid in which one atom is located at each vertex of a cube. Much more complicated, but periodic, arrangements of atoms are possible in a crystal. The atoms are held in place by electromagnetic forces and the typical separation between the atoms is a few Angstroms.

What keeps these atoms in such a regular array in a solid? Incidentally, it is the electromagnetic forces which determine the overall structure of matter and the force between atoms in a crystal lattice is of electrical nature. Even though individual atoms are electrically neutral, the currents produced by the orbiting electrons inside the atom exert a force on the nearby lattice atom. This force is the cause of the binding energy of a crystal lattice. You know from experience that it takes fair amount of effort with an ice pick to break an ice cube into smaller pieces. In this process, one has overcome the forces that hold the crystal together by supplying energy to the block of ice.

What will happen if we keep supplying more and more energy to a solid body? At first the atoms in the crystal lattice will vibrate with larger and larger amplitude and soon — if we supply sufficient amount of heat energy to a block of ice — it will melt and become water. Heating provides enough energy to break the binding between the atoms in a crystal lattice, thereby overcoming the rigidity of the ice. The same phenomenon takes place in any other solid. As we heat a solid — and supply more and more energy — there occurs a moment when the solid melts and forms a liquid. Different solids, of course, will require different amount of heat energy to be supplied for their melting; it is comparatively easier to melt an ice cube than a piece of iron. It takes about 80 calories of energy to melt one gram of ice.

This quantity, "calorie", is a unit for measuring the energy. (It is related to the unit you keep hearing in connection with exercise and diets, all of which have to do with supplying or burning of energy.) In physics, it is also usual to use several other units for energy, of which the most frequently encountered one is "erg". One erg is about 2.38×10^{-8} calories; so we may say that it takes 3.4×10^9 ergs to melt one gram of ice. The rigidity of ice is then lost and water "flows" easily.

We can, of course, continue the above experiment by supplying more and more heat energy to the liquid. At first, the water will just get hotter and hotter. Soon, another change will occur and the liquid will become a gas. That is, water will change to steam. You must have noticed that steam is nearly invisible, compared to water. This is because the atoms forming the steam are lot more sparsely distributed compared to the atoms in water. [One cubic centimeter of steam will have nearly a thousand times less atoms than one cubic centimeter of water]. The gaseous phase is also common to any other solid; by supplying sufficient energy to molten iron we can produce a gaseous vapour of iron atoms.

This description covers three of the most familiar states of matter: solid, liquid and gas. In nature, there is also one more phase which occurs at sufficiently high temperature, called the plasma state. When you supply sufficient amount of energy, one can rip apart the electrons from the atoms leading to positively charged ions and negatively charged electrons both of which exists in a state of random motion. This, of course, involves some amount of atomic physics and hence will become clearer when we describe the structure of the atom in Chapter 3.

1.5 Entropy: A new beast

In the last section we described how the behaviour of bulk matter, like a gas of hydrogen atoms or a crystal lattice, can ultimately be understood in terms of the laws of microscopic physics. We illustrated this by showing that the ideal gas law which governs the behaviour of gaseous systems can be obtained by applying the simple laws of Newtonian mechanics to the molecules which are randomly moving in a volume of space. Similar considerations apply to more complex systems and one can always obtain the behaviour of macroscopic systems from the laws of microscopic systems.

This fact suggests that macroscopic bodies do not bring in any new phenomena which are not inherently present at the microscopic level. While this is true at some fundamental level, it is not of much practical value in studying the physics of macroscopic systems. Very often, it is convenient to introduce concepts in macroscopic systems which have no interpretation at the microscopic level. Some of these concepts, as it will turn out, play a crucial role in our discussions in later chapters. In view of this, we shall take a closer look at such concepts before proceeding further.

To illustrate the ideas in a simple context, let us again consider a gas made of, say, hydrogen molecules confined in a large volume V. For any macroscopic volume, the number of molecules contained in it will be enormous; for example, in one litre of hydrogen gas under normal room temperature and pressure, there will be about 10^{22} molecules. Each of these molecules will be moving randomly with some average speed. The temperature of the gas — as we described in the previous section — is essentially a measure of the average kinetic energy associated with the random motion of these molecules. Obviously, one needs a large number of molecules in order to define a useful notion of an average. This is, of course, true in any situation in which one wants to introduce the concept of an average. For example, to compute the average monthly income of the people in a city we add up the total income of the people in the city and divide it by the number of people living there. This makes sense only when the number of people in the city is sufficiently large. If there are only 2 or 3 persons living in a hypothetical city, you can still compute the average but it may not mean much if the income of the 3 people differs widely. In a similar manner, the concept of temperature in a gas makes sense only when large number of molecules are involved and there is no such thing as a temperature of an individual molecule. Therefore, temperature is the first example of a physical concept which makes practical sense when large number of particles are involved but cannot be defined usefully at the microscopic, molecular level.

Similar comments apply to even the idea of density of a gas — or even that of a solid. When we say that the density of iron is 7.87 gm/cm^3 we refer to a piece of iron containing a large number of iron atoms. Most of the region inside the solid however is empty and — in fact — even most of the region inside the atom is empty. The mass of an atom is entirely concentrated on the nucleus which has an extraordinarily high value of density. So we again see that the notion of 7.87 gm/cm^3 as the density of iron is a completely macroscopic concept and cannot be defined usefully at the microscopic level.

These two examples illustrate the point but they are, in some sense, easy to understand and not surprising. We shall next introduce a concept called *entropy* which plays an important role in virtually all areas of physics and has some counterintuitive aspects associated with it. Instead of giving a mathematical description, we shall try to introduce it in terms of simple examples.

To do this we need to bring in the concepts of microscopic state (or *microstates*, for short) and the macroscopic state (or *macrostate*, for short). Consider an array of N light bulbs located in a line each of which can be switched on or off. Suppose you switch on a set of them, say, the 1st, 3rd, 17th, 123rd .... and tell me which of them are turned on and which of them are off. This completely specifies the microstate of the system in this case. You now know precisely which bulb is on and which bulb is off. Each of the bulbs can exists in two different states (on or off) and hence the total number of possible states of the system is $2 \times 2 \times ...N$ times $= 2^N$. This is a very large number even for reasonable N. For example, for $N = 270$, the total number of possible states for these bulbs is larger than the number of atoms in the observed universe.

Let us now consider a slightly different situation. You again switch on certain number of bulbs (say, k bulbs) but do not tell me which bulbs are switched on. (Alternatively, you can imagine that I measure the total amount of illumination produced by the array of bulbs from some distance and figure out that k bulbs must be switched on; but I have no way of accurately determining *which* k bulbs are switched on.). I now have less information compared to the previous situation in which you told me the exact state of each bulb. The information I now have — viz. that some k bulbs are switched on — specifies the macrostate of the system. The key point is that, there are large number of microstates which are consistent with the information I have about the macrostate. Suppose $k = 2$; that is I only know that 2 bulbs out of N are switched on. The first of these could be any one of the N bulbs. Having fixed that, for each of the choices, the second bulb could be any of the $(N - 1)$. So there are $N(N - 1)$ possible ways of doing this. But the final state of the system in which I first switch on 3rd bulb and then switch on 17th bulb is the same as the one obtained by first switching on the 17th bulb and then switching on the 3rd. So only half of the $N(N-1)$ possibilities are really different. We thus conclude that there are $(1/2)N(N - 1)$ distinct microstates of the system which lead to the same marcostate in which two bulbs are switched on. The macrostate contains considerably less information than the microstate.

This lack of information is described by a quantity called entropy. If a given macrostate can arise due to 2^S (that is 2 multipied S times) mi-crostates, then S is called the entropy of that macrostate. These concepts can be applied to any system with large number of subcomponents (like the bulbs in the above example). For example, in a gas made of N par-ticles, the microstate is specified by the positions and velocities of each of the molecules. But usually we specify the state of such a gas by giving its total energy E and the volume V it occupies. There are ways of com-puting the total number of possible microstates (positions and velocities of the molecules) which can lead to the same E, V and from this one can determine the entropy of a gas S as a function of its energy and volume. It turns out that once we know this, one can determine all other physical properties of the gas.

There is something very curious about the behaviour of entropy in any physical system which makes it such a useful and important concept. It turns out that when macroscopic systems interact with each other, the entropy always increases or, at best, stays constant. You have to arrange things in a special manner and spend energy in order to decrease the entropy of any given system. Even when you do that, it will turn out that the entropy of some other part of the external environment, which is coupled to the system we are working with, goes up. When you calculate the net change in the entropy of the external environment as well as the system we are interested in, it will turn out that the total entropy has increased. This result — that the entropy keeps increasing during physical interaction — is called the second law of thermodynamics and is one of the key principles of macroscopic physics.

To see how this comes about, let us consider a simple example shown in Fig. 1.15. The part (a) of the figure shows a box of volume V which is separated into two halves by a partion in the middle. On the left half, we have a large number of molecules of a gas at some temperature and pres-sure in random motion while the right half is initially empty. This system has some amount of entropy which, of course, will depend on the volume of space which is available to each molecule. In any given microstate of the system, each of the molecules could be found anywhere in the volume $V/2$ and just specifying the macrostate by giving the pressure, temperature and the volume of the region occupied by the gas does not give any further infor-mation about the microstate. Let us suppose we now remove the partition at a given instant of time. At the next moment, we will have a situation as described in Fig. 1.15(b). All the molecules are still in the left half of the volume but, of course, they are now free to move throughout the volume. In fact, this is precisely what will happen if we wait for a little while. The gas will expand and very soon the molecules can be found anywhere in the volume V. Each of the molecules can now be found within a larger volume compared to the situation shown in Fig. 1.15(b). Clearly, the number of possible microstates available to the system has gone up in the process and hence the entropy of the gas has gone up.

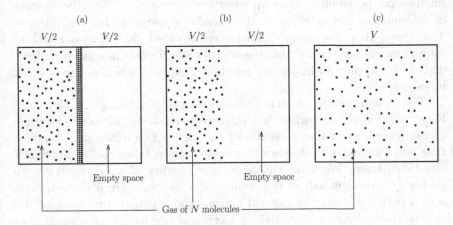

Fig. 1.15 Gas molecules, which were initially confined to the left half of a volume of a box (in Part (a)), are allowed to spread to the full volume by removing the partition. The situation just after the partition is removed is shown in Part (b) and the situation at a much later time is shown in Part (c).

There are two points that need to be stressed regarding the above — relatively simple but instructive — example. First, there is a connection between the entropy and our lack of information about the systems. Let us say that there were k possible microstates available to the system when the gas is confined to the left half of the volume. Our ignorance about which microstate the gas is actually in will be mathematically described by the entropy associated with this state. When the partition is removed and the gas fills the full volume of the box, the total number of available microstates has gone up to, say, k' with $k' > k$. This is what has led to the increase in entropy. But we also have less information about the system now. Originally we knew that the gas is in one of the k possible states, but now the gas is in one of the k' possible states. Since $k' > k$, our information about the system has actually come down.

If you have difficulty in grasping this concept, just compare it with a common situation. Suppose you know that your friend has an office in the 14th floor of a building which has 20 floors with each floor having 10 offices. Then you know that she is in one of the 10 offices in the 14th floor. But suppose you only know that she works in that building, but do not know which floor her office is. Now she could be in any of the $20 \times 10 = 200$ offices. You will certainly agree that when the number of microscopic possibilities goes up from $k = 10$ to $k' = 200$, the amount of information you have about the friend's whereabout has come down. This shows that the entropy of a system is closely linked to our lack of information about the different possibilities which exists in a given context. This idea will play an important role in a Chapter 6 when we study black holes.

The second feature which is immediately evident from Fig. 1.15(b) and Fig. 1.15(c) is the following. Suppose we were shown these two pictures as the snapshots of the molecules of the gas at two different instants of time and were asked to decide which snapshot was taken earlier and which was taken later. We know that if a large number of gas molecules were confined to the left half of the volume of a box at a given instant, then over a period of time the gas will expand and fill the entire volume. So, clearly, the snapshot in Fig. 1.15(c) was taken at a later time compared to the one in Fig. 1.15(b). In other words, the direction of increase of entropy in some sense allows us to determine the direction of flow of time. One usually says that the above example provides a thermodynamic arrow of time. So, entropy is connected in a mysterious way with the deep question of what determines the irreversible flow of time which we all experience.

The existence of such an arrow of time raises another interesting question. All the fundamental laws of physics — like, for example, the laws of mechanics — remain the same if we change the sign of time; that is, if we interchange past and the future. Suppose we throw a ball up in the air and take snapshots of the trajectory of the ball during a given time interval. If you are shown the sequence of snapshots, there is no way you could have determined in which sequence the snaps were taken. This is because, for any given trajectory of the ball, there exists another equally valid trajectory in which the ball passes through the same spatial location in the reverse order. What is true for a ball is also true for any single molecule which is located on the left half in Fig. 1.15(b). If you were only given the trajectory of any given molecule at different instants in time, you cannot arrive at an arrow of time from it. For any given trajectory there is another possible trajectory which is obtained by reversing the direction of the time. In spite of this, if you are given the behaviour of very large number of molecules — as in the case of Fig. 1.15(b) and Fig. 1.15(c), one can immediately determine the direction of time.

That is definitely peculiar and people have investigated this thoroughly. The crucial question one needs to answer is the following: Suppose we start with the situation shown in Fig. 1.15(c) with the molecules moving randomly in all directions. Since the motion is random, isn't it possible that after sometime all the molecules end up simultaneously in the left half of the volume thereby reaching the configuration shown in Fig. 1.15(b). Since the laws of microscopic physics are invariant under time reversal, this is certainly possible, in principle. All we need to do is to start with a situation in which the positions of the molecules are as in Fig. 1.15(c) with the direction of motion reversed; then the molecules will end up at the configuration Fig. 1.15(b) after some time. The right question to ask, however, is how probable is it? From everyday experience, we know that gas molecules do not behave like that. (Imagine what would happen if all the oxygen molecules in your room decide to move away from you and stay in an opposite corner for a few minutes!) If there are N molecules in the gas and the motion is truly random, any given molecule has an equal chance to be found either in the left half or in the right half of the box. So the probability that any given molecule is in the left half is $(1/2)$. The chance that all the N molecules end up in the left half simultaneously at any given instant is $(1/2) \times (1/2) \times \cdots N$ times $= 1/2^N$. For any reasonable N, this is an extremely tiny number and hence the probability of it occurring is utterly negligible. If you take one litre of gas, one can be quite confident that such an event is not going to take place within the entire lifetime of the universe and in fact our intuition — for once, correctly — has grasped that, for all practical purposes, we can say that a gaseous system will not proceed from the state described in Fig. 1.15(c) to the state described in the Fig. 1.15(b). In short, even though such a transition is allowed by the laws of physics, it is an event of such a low probability that we can assume that it does not occur.

If the molecules in the gas proceed spontaneously from the state in Fig. 1.15(c) to the one in Fig. 1.15(b), then the entropy of the gas will decrease during this process. Our conclusion that such an event does not take place *for all practical purposes* shows that the entropy of the gas does not decrease spontaneously for all practical purpose. You see that the laws of thermodynamics have brought in certain probabilistic consideration into physics unlike the fundamental laws of microscopic physics. The increase of entropy during physical processes occurs with such an extremely high probability that one can take it to be a certainty.

1.6   A peep ahead: Where did it go wrong?

It turns out that the ideas of statistical mechanics has far reaching impact and — with future use in mind — we shall describe some of these. To begin with, let us take a closer look at the relation between the kinetic energy K and the temperature T given by $K = (1/2)mv^2 = (3/2)k_BT$. A gas molecule can move in three different directions in space and one says that the molecule has three degrees of freedom. The above relation shows that one can associate a thermal energy $(1/2)k_BT$ with each degree of freedom. This, of course, only takes care of the translational kinetic energy arising due to the entire molecule moving as a unit. If the gas is made of diatomic molecules, say, then there can be other sources of energy. For example, the two atoms of a diatomic molecule can vibrate along the axis joining them and the vibrational mode stores some amount of energy. This adds another degree of freedom in which energy can be stored. A diatomic molecule can also rotate in the plane of the two atoms which allows for yet another way of storing energy. A more complex molecule can vibrate and rotate in more complicated ways so that it will have more degrees of freedom. A key result in classical statistical mechanics, as formulated by Boltzmann, is that when a system is in thermal equilibrium, each degree of freedom can store $(1/2)k_BT$ of energy. So if we have a gas with N molecules with f degrees of freedom per molecule, then it will have an energy $(f/2)Nk_BT$. In other words, to increase its temperature by one degree, we need to supply an additional energy of $(f/2)Nk_B$. This quantity — the energy that needs to be supplied to increase the temperature by unity — is called the *specific heat* of the system and can be directly measured. We now have a clear prediction about the specific heat of different kinds of gases (monoatomic, diatomic etc.) and one can compare the predictions with observations.

When this was done, the results were surprising. One found that, at sufficiently high temperatures, the results agreed with observations while at lower temperature there were serious discrepancies between the observations and theory. It was as though some of the degrees of freedom were not participating in the energy sharing process when the temperature was low. This was one of the first indications that the energy of microscopic systems, like molecules, do not behave in the expected manner. We will see in a Chapter 3 that this discrepancy was actually one of the factors that led to the quantum revolution.

There was another — and more striking — discrepancy which statistical mechanics brought into classical physics. To understand this, we need to first introduce the concept of thermal radiation. This is the electromagnetic radiation emitted by any hot body due to its temperature. You know that when an iron rod is heated to a high temperature it emits light (which is a form of the electromagnetic radiation) of predominantly red colour; we say that the iron is "red hot". This is merely a particular case of a very general phenomenon. When any material object is kept at some nonzero temperature, it emits radiation of a particular characteristic which depends only on the temperature.

A simpler way of thinking about this radiation is as follows: Consider a closed metallic box, the surface of which is kept at some fixed temperature. Then the inside of the box will be filled with electromagnetic radiation with a characteristic pattern in energy density which depends only on the temperature of the box. More precisely, the amount of radiation energy which is in the box at a given frequency can be expressed entirely in terms of the temperature of the box. If you put a small hole on the box and let a little bit of radiation to leak out, we can figure out how much of radiation energy is there at any given frequency. If you make two boxes of different shapes, sizes and made of different metals and keep them at the same temperature, then the radiant energy inside the boxes will have identical form. The question arises as to whether we can understand this phenomenon from theoretical considerations.

We saw earlier that electromagnetic radiation is essentially due to oscillations of electric and magnetic fields and these oscillations can be thought of as different degrees of freedom associated with the oscillating electromagnetic field. Each mode of vibration is capable of storing energy. Using our rule that each degree of freedom should store an energy $(1/2)k_BT$ and counting the number of degrees of freedom associated with a vibration at frequency ν, one can compute the amount of radiation at a given frequency. Such a calculation, originally done by Lord Rayleigh led to the result that, the amount of energy U at a frequency ν should vary as $U \propto \nu^2T$ at temperature T. This was a rigorously derived result from theory; the trouble is, it is an absurd result!

To see its absurdity, you need to note that the result claims that the energy increases as square of the frequency at any given temperature. So if you keep a metallic box at room temperature and peep into it, most of the radiation that hits you will have the highest frequencies; lots of X-rays, gamma rays etc. This is manifestly wrong. Further, the results show that any box at any nonzero temperature has an infinite amount of total energy! If you add energies at all the different frequencies you get an arbitrarily large value — mathematically speaking, infinite value — because there is more and more energy at higher and higher frequencies.

While the theoreticians were grappling with these absurdities, the experimentalists were also at work. They found out how exactly a box at some temperature behaves. Of course, nothing strange was happening in real life in which the radiation inside the box had the form in Fig. 1.16. The radiation does contain a wide band of frequencies but observations show that it will be peaked at some frequency with very little radiation at much higher frequencies. The frequency at which one has maximum radiation, increases with temperature of the box. Correspondingly, the wavelength of the peak decreases with the temperature. Bodies at room temperature (which corresponds to 300 K) will emit most of the radiation in the infrared band. If the temperature is increased to about 10^4 K, the radiation will be mostly in the visible range and if we heat the system to about 10^8 K there will be copious production of X-rays. The amount of radiation emitted at different frequencies by a body at some temperature, is called "thermal spectrum" or "Planck spectrum".

At low frequencies the law $U \propto \nu^2 T$ discovered by Lord Rayleigh was indeed true; the energy keeps increasing with frequency up to a point; but then it starts to decrease rapidly at higher frequencies. At any given temperature the maximum amount of energy was at some frequency which rises in proportion with the temperature. The total amount of energy in the box was, again, finite and increased as T^4. If you doubled the temperature, the total radiant energy went up by a factor 16 etc. Clearly, the theory falls in its face trying to get the results right!

So what is going wrong ? Again the result has to do with quantum theory which we will take up in Chapter 3 but the key point is the following. The results described above assume that the radiation is in equilibrium with matter which makes up the metallic box — that is, the amount of energy the box was absorbing from the radiation should be equal to the amount of energy the box was emitting into the box. In doing the calculation it was assumed that the matter which makes up the box can exchange energy with the radiation by arbitrary amounts. Remember that if the box is kept at some temperature, it will make the atoms of which the box is

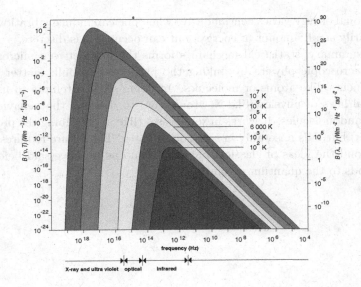

Fig. 1.16 Any object kept at some temperature emits a characteristic spectrum of radiation called the Planck spectrum. As the temperature is increased, most of the radiation will be emitted at shorter and shorter wavelength. Note that, in the figure, the frequency *decreases* to the right. The figure shows the nature of emission for bodies kept at temperatures ranging from 100 K to 10^7 K. Bodies at room temperature (300 K) emit most of the radiation in the infrared. Bodies at higher temperatures, say, between 5000 K to 7000 K (as in stars) emit large amount of visible radiation.

made of, vibrate in a characteristic manner. Assuming that the energy of these vibrations can take all possible values is equivalent to saying that the box can exchange energy with radiation by arbitrary amounts. Quantum theory changes this result and leads to the conclusion that the energy of vibration of the atoms in the box is quantized. This means the radiation can be exchanged only in quantized units. When you redo the calculation with this assumption, you get the correct result; the theory reproduces the experimental result.

It turns out that the same assumption — that the energy is quantized — allows one to understand the strange behaviour of the specific heat as well. At low temperatures, there is not sufficient energy to excite the modes of vibration and hence lots of degrees of freedom do not participate in sharing of energy or contributing to the specific heat. This cannot arise if the vibrational energy varies continuously (when one can induce vibration with arbitrarily small amount of energy) but can occur if it is discrete.

In summary, statistical mechanics forms the link between microscopic and macroscopic physics by linking the properties of bulk matter to its constituents, like atoms or molecules. In doing so, it attempts to use the classical laws of physics (like Newton's law) to describe the behaviour of atoms and molecules. It works in some cases (like in describing the pressure of a gas) but fails miserably in some other cases. The failure requires us to go beyond the ideas of classical physics — and as we have said before — this leads to the quantum revolution.

Chapter 2

And then there was Einstein

Thus, at the end of 19th century most physicists felt that they knew all the fundamental laws which govern the behaviour of nature and that they just need to do the calculations more accurately to understand everything. Amidst this complacency there was a prophetic comment from Lord Kelvin. In a Friday evening lecture at the Royal Institution delivered in April 1900, Kelvin stressed that while we do seem to understand most of the physical phenomena in nature, the "beauty and clearness of the dynamical theory" is "at present obscured by two clouds". The "clouds" he was referring to dealt with the the difficulties in understanding the propagation of light and the specific heats of gases.

With hindsight, one cannot but appreciate his genius. At any given time, physicists will be grappling with several unsolved issues and it is never very clear which of them are fundamental and which of them are incidental issues pertaining to details. A great physicist is the one who can concentrate on the deepest of the problems and Kelvin certainly identified the right ones at the end of 19th century. When these two clouds rained, the first one led to relativity (which is the subject of this chapter) and the other led to quantum theory (which we will discuss in the next chapter) thereby heralding two major revolutions of the 20th century physics. Let us begin with relativity.

2.1 Something crazy about light

Let us start with the thought experiment similar to the one we briefly discussed in the last chapter. You are sitting inside a car moving forward with a steady speed of, say, $v = 60$ km/hr. You now throw a ball towards the front of the car with a speed $u = 10$ km/hr *with respect to the moving car.* We saw in the last chapter that the ball will acquire the instantaneous speed of the car; therefore, you friend standing by the side of the road will see the ball moving forward with a speed of $(v + u) = (60 + 10)$ km/hr $=$ 70 km/hr with respect to the ground. What is true for the ball must be true for anything else, right ? Well, not quite.

The trouble arises when instead of throwing a ball you emit a beam of light. Suppose you switch on a torch beam towards the front of the car, when the car is moving forward with the speed v. Light normally travels with an incredible speed of $c = 300,000$ km/sec. If the light is to behave like a ball, then your friend on the ground will see it travel forward with a speed of $(c + v)$. Similarly, if you shine a beam of light towards the back of the car (that is, in the direction opposite to the motion of the car), then it will travel with a speed of $(c - v)$. Thus, in any moving car the headlight beam should be travelling with a speed $(c + v)$ and the rear light beam with a speed $(c - v)$.

During the last part of 19th century, people attempted to test this out by direct experimentation. The actual experiment is fairly complicated but the rough idea is to use the fact that the Earth itself is moving in space. If you bounce a beam of light between two mirrors with its path parallel to the direction of motion of Earth, then its speed should be different during the forward motion and return trip. By sensitive techniques, one can measure this difference very precisely. All such experiments which attempted to measure this gave a null result. Light simply did not care how fast the source was moving. The headlights and the tail lights of a moving car propagate with a speed precisely equal to c; not with $(c + v)$ and $(c - v)$! In this respect, the light was behaving very differently from other material bodies.

At this stage you might say, "Big deal! We all know light is different from a plastic ball, so how does it matter if it does not share the speed of the source?" But actually, it is a "big deal". The point is that the rule for addition of speeds $(c + v)$ is deeply linked to the way we measure time and distances in a moving car compared to the stationary road. If the $(c + v)$ rule fails, then our ideas about measuring spatial positions and time intervals will go for a toss. All hell will break loose and it did. The result is called Special Relativity.

2.2 Time and its flow

Let me begin by describing one serious consequence of what we have just seen: viz., that the speed of light remains constant independent of the speed of motion of the source. Consider a long railway cabin shown in Fig. 2.1. In the middle of the cabin (B in the figure), there is a source of light which can emit beams of light towards the two ends of the cabin, A and C. The train is moving with C as the forward end. Let a physicist Joe ride in the train while another physicist Moe stay on the road watching what happens. We have decided that the speed of the light is going to be a constant c as observed by either of these two observers. With respect to Joe, the train is not moving at all. Further, since the distances AB and BC are equal, the light beams emitted from C and moving with speed c towards A and B will hit A and B simultaneously. Now, according to the Galilean principle of relativity we described in the last chapter, the physical phenomena as described by Joe should be identical to the physical phenomena described by Moe in the stationary frame. So we would expect Moe also to see the light beams hit A and C simultaneously.

Let us now study this from the point of view of Moe. From his point of view, the end A of the train is moving towards the light beam while the end C of the train is moving away from the light beam. Obviously the light beam travelling towards A will have to cover less distance than the light beam travelling towards C. Since the observations show that the two beams are travelling with the same speed — and this is the crucial point — it will take less time for the beam to hit A than for it to hit C. So Moe will see the beam hit A first and, a little later will see it hit B. The two events (the light beam hitting A and the light beam hitting C) are not simultaneous as far as Moe is concerned while Joe will insist that these two events are simultaneous. *The notion of simultaneity is not absolute but relative!*

Note that this would not have happened if the speed of light changed according to our old rule. In that context the speed of light beam travelling towards C would have been enhanced to $(c+v)$ (as far as Moe is concerned) and the speed of the light beam travelling towards A would have been reduced to $(c-v)$. This will precisely compensate for the fact that one beam has to travel more distance than the other. Both observers will see the events as simultaneous and everything will be quite sane.

But that is not to be. Experiments convincingly show that the speed of light does not change due to the motion of the source. So we reach the inescapable conclusion described above that the notion of simultaneity

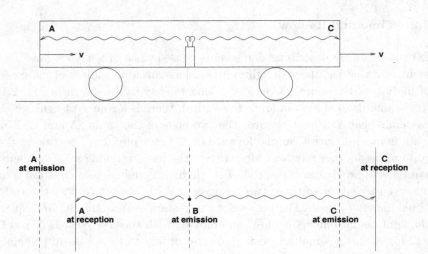

Fig. 2.1   Thought experiment illustrating why two events which occur simultaneously according to one person need not occur simultaneously according to another person. The top frame shows a rail carriage moving to the right, with a light source in the middle. The source emits two beams of light towards the front and back ends of the rail carriage. According to an observer travelling in the carriage, the beams hit the two ends simultaneously. The bottom frame shows the result from the perspective of a stationary observer, relative to whom, one beam has to travel a larger distance than the other. Since experiments show that beams of light always travel with the same speed, this observer will see the beam hit the back end before the front end.

depends on the observer. As you would have noticed, it is plain logic and requires nothing more than high school algebra. (In fact, this is true of most of the Special Theory of Relativity; the ideas are subtle but the mathematics, surprisingly, is simple.)

You might think it is a bit strange but still not really earth shaking. What if simultaneity is not an absolute concept? The point is that if simultaneity is not absolute, the flow of time cannot be absolute either. One way to see this result — somewhat abstractly — is to ask what we really mean when we say "I started having breakfast at 08:00 hours". What we mean is that there are separate events, namely (i) me starting to eat my cereal, and (ii) the hands in my wrist watch pointing towards the digits 8 and 12 are *simultaneous*. If simultaneity is a relative concept, then assigning a definite time to an event becomes relative. In other words, the flow of time is no longer absolute.

Fortunately, there is a very concrete way of demonstrating this, once again without any complicated mathematics. To do this, we will build (of course, all in our imagination) a simple kind of clock using light rays. This is shown in Fig. 2.2(a). At the bottom is a light source A which emits a pulse of light upwards to B where a mirror reflects the beam back to the bottom. (For clarity we have shown the point at which the light returns back, C, slightly separated from A.) The time taken for the light to go up and come down is $2l/c$ where l is the vertical distance between the source and the mirror. If a pulse of light is sent out periodically between A and B, then by counting the number of times it bounces back and forth we can measure time. If it bounces 10 times, say, the total time interval is just $10 \times (2l/c)$. You need to keep counting to use this clock, making it quite impractical, but there is no question that theoretically this is a device to measure flow of time.

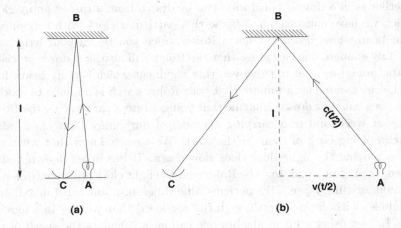

(a)   (b)

Fig. 2.2   Moving clocks run slower than stationary clocks, making flow of time relative. Left frame shows a 'light-clock' which involves a gadget (A) which emits light, that is reflected by a mirror B to a receiver C. The time taken for the round trip allows one to measure time. Right frame shows what happens to this clock when it is moving horizontally to the left. The light now has to travel a longer path but since it travels with the same speed, the rate of flow is affected.

Let us now suppose that someone takes this clock and moves with it at a uniform speed v along the horizontal direction. For an observer moving with the clock, nothing has changed and she will measure a time interval $t_{mov} = 2l/c$. For the stationary observer, the situation now is as described in Fig. 2.2(b). Between the time light goes from the bottom to the top, the mirror has moved forward a little and again between the time the light is reflected from the mirror and reaches back at the ground, the receiver at the bottom has moved a little further (see Fig. 2.2). It is clear from the figure that the light now has to travel a longer distance than $2l$ but since light travels with the same speed in all frames (here we go again!), it will take a longer time for it to bounce back and forth.[1] The time interval between events as measured by the moving clock is shorter compared to the time interval as measured by a stationary clock. When the moving clock shows, say, 1 hour duration the stationary clock will show a longer duration; say, 1 hour and 10 minutes. In everyday language, we say that the moving clocks run slower. Since flow of time is what we measure with clocks, physicists have no hesitation in saying that the flow of time is relative and not absolute!

That, of course, should have shaken you to the core. You may wonder whether such a drastic conclusion can be drawn from a rather funny clock which we have constructed. Maybe this particular clock with a bouncing light beam slows down but not a Rolex which you tie on your wrist. Incredibly enough, one can *prove* that the Rolex will also slow down precisely at the same rate. Let me give you this argument which is truly beautiful.

Let us assume for a minute that your Rolex watch is immune to motion and does not slow down. Imagine that you get into a car wearing the Rolex on your wrist and also carrying a model of our funny light clock, after synchronizing both of them at the start. We already know that when the car is moving the light clock does slow down. If the Rolex does not slow down, then after some time the Rolex and the light clock will show different amount of time lapse. By noticing this difference you can immediately conclude (without looking through the windows) that you are in a moving car. In fact using a bit of algebra one can even calculate the speed of the car. But this is precisely what the Galilean relativity forbids us from being able to do! We should not be able to decide whether the car is moving with uniform speed or standing still by performing experiments inside the car. This requires the Rolex to slow down exactly the same way as our light clock does!

In fact, the two principles which we take as correct, viz. (a) physical experiments inside a moving car will not allow us to determine whether it is moving with uniform speed or at rest and (b) the speed of light is a constant

---

[1]If you remember your Pythagoras theorem, you can immediately obtain from Fig. 2.2(b) that $(vt/2)^2 + l^2 = (ct/2)^2$, which gives $t\sqrt{[1 - (v/c)^2]} = (2l/c) = t_{mov}$. So the time interval as measured in the moving frame t_{mov} is shorter than the time t measured in the frame of stationary observer.

in all frames of reference, imply that *all natural phenomena* must slow down in a moving frame. Suppose, for example, we have a bacterial culture which doubles its population precisely in every 10 seconds. In principle we can use it as a clock. We now take the bacterial culture and our light clock in a moving car. If the growth of bacteria does not change, then by comparing it with the light clock we will immediately know that we are moving; since this is forbidden, the growth of bacterial culture must slow down when it is moving! Your own biological processes will slow down and you will age less, when you are in motion!

Fortunately, there is a very direct experimental verification of the fact that things live longer when they are in motion. There is a particle called muon (which is very similar to electron except that it is more massive; we will discuss this in more detail in Chapter 4) which is unstable and decays to less massive particles in a very short period of time. The mean lifetime of a muon is about 2.2 micro second. Even if the muon travels with the speed of light it can only travel a distance of about 660 metres during its life time. It has been found experimentally that the muon actually travels over much larger distances before decay. The reason essentially is what we saw above. The muon travels with speeds close to that of light and hence, in its reference frame, time flows more slowly and it lasts longer. This has been a very real effect.

The result obtained above is probably one of the most powerful results obtained in physics, using two innocuous looking assumptions and a clever argument. Since the conclusion looks strange, people have tried to get around it in all possible ways but no-one has succeeded. Every experiment which is performed to test these ideas has given verdict in their favour.

2.3 Relativity in action

It is obvious that the inescapable conclusion reached above, viz. the flow of time is relative and is different for different observers, is going to have far reaching consequences. In a way the entire special relativity revolves around figuring out the different consequences of this result and reformulating the old ideas correctly. Let us look at some of them.

The fundamental concept in the description of special relativity is that of an *event* which — fortunately — means exactly what the plain English language indicates. An event takes place at some point in space at a given time and can be specified by giving its location in space (which will need three numbers) and the time (which will require one number). So an event is characterized by four numbers and mathematicians would like to say that an event is a point in a four dimensional space or, more precisely, in four dimensional spacetime. This is really no different from us calling the surface of a sheet of paper two dimensional, because we need two numbers for specifying the position of any point on the paper, one along the length and one along the breadth. Sometimes one makes it all sound very mysterious and philosophical by talking about time as the "fourth dimension". This is not very helpful and ultimately what matters is that each event we are interested in requires four numbers to characterize it rather than three — which is a fact that anyone can understand.

In the Newtonian world, the time interval between any two events is the same for all observers though the spatial distance can be different if one observer is moving with respect to the other. In relativistic description, both the time intervals and the spatial intervals between any two events are different for different observers. So, in special relativity, different observers will attribute different sets of four numbers to the same event. That two observers will allot different spatial positions to different events is obvious and exists even in Newtonian mechanics; but in Newtonian mechanics, time is absolute and all observers will assign the same time coordinate to any event. But as we saw in Sec. 2.2, the rate of flow of time is different for different observers in special relativity and hence different observers will assign different values of time and space for the same event. The relation between the time and space coordinates attributed to a given event by different observers can be expressed by a formula called *Lorentz transformation*. This formula forms the cornerstone of special relativity and all other relevant consequences can be obtained from it. We will now describe some of them.

Let us get back to the starting point of the discussion — which was the addition of speeds when an object is thrown inside a moving vehicle. If a car is moving with speed v and you throw a ball sitting inside the car with a speed u relative to the moving car, we thought that the ball will move with a speed $u' = v + u$ with respect to the ground. This landed us in all kinds of difficulties with experimental results when the ball was replaced by a light beam. The correct formula (for any object, plastic ball or light beam) turns out to be different but still quite simple. It is given by $u' = (u+v)/[1+(uv/c^2)]$. where c is the speed of light. The first thing you notice is that the quantity in the denominator is essentially unity for even the fastest vehicles on Earth. If you shoot a bullet with speed $u = 100$ m/s while travelling in a superfast train going at $v = 300$ km/hr $= 83$ m/s then the quantity uv/c^2 is just one part in 10^{13}. So, far all practical purposes, the above equation can be thought of as same as $u' \approx u + v$. This is the reason we don't have to worry about relativity in our everyday life. The speeds we are interested in are usually much smaller than the speed of light and the kind of corrections to our usual formulas are negligible at these low speeds.

But consider now what happens to the headlight beam of the car moving with a speed $v = 100$ km/hr. Since the light beam travels with the speed of light, $u = c$ and the denominator becomes $1 + (v/c) = (c + v)/c$; the numerator is $c + v$. So we get the beautiful result that $u' = c$! That is, if you emit a beam of light from a moving car, it still moves with the speed c according to our formula, independent of the speed with which the car is moving. On the other hand our non-relativistic formula would have given the (wrong) result $u' = v + c$. The correct formula is designed to make sure that the speed of light remains constant irrespective of the speed of the source.

The above result also tells you that you cannot accelerate a material body to speeds greater than that of light. One way to understand this result is to think of acceleration as incrementing the speed of the body step-by-step. If a body was moving with a speed v and you give it a small extra speed q we would have thought that its new speed will be $v + q$. By repeating this process we can increase the speed to $v + q, v + 2q, v + 3q, ...$ etc. without limit. Obviously at some stage the speed will exceed c, the speed of light. But our new analysis has taught us that the simple addition is incorrect. As you get closer and closer to the speed of light, the factor in the denominator, $[1 + (uv/c^2)]$, becomes larger and reduces the effectiveness of our acceleration. At best, you can asymptotically approach the speed of light c but can never exceed it. The speed of light thus acts as a barrier to material bodies. Nothing can be made to travel faster than light by accelerating it. In fact, one can prove a more general result in relativity that no signal can propagate with a speed greater than that of light.

This raises another related question. In Newtonian mechanics we saw that a force exerted on a body increases its momentum $p = mv$. If this is true, then by applying a steady force to a body, we should be able to increase the momentum to any large value we want and thus increase the speed to any large value we want. But we just saw that this is impossible according to relativity. Something has to change again. This time what gets modified is the relation between momentum and velocity and the correct result makes sure that a steady force acting for a long time will only make the particle acquire a speed close to that of light but never exceed it.

One can also look at the same phenomena in terms of the energy of the particle. When you push a particle to make it move, the work done by you appears as the kinetic energy of the particle. In Newtonian mechanics, the kinetic energy has the formula $E = (1/2)mv^2$. If this is correct, then by doing a steady amount of work over a long period of time one should be able to increase the kinetic energy to as high a value as possible and thus the speed should eventually exceed c. Once again, since this is forbidden in relativity, the formula for the energy of the particle in terms of its speed should get modified. Einstein derived the correct formula using relativity and found that it does take care of this difficulty. In the correct formula, as the speed approaches the speed of light the total energy will increase without bound. In other words, you have to do infinite amount of work to make a particle reach the speed of light; with any finite amount of work, you can only get the speed arbitrarily close to c but not to exceed it.

The formula for energy which Einstein derived has another remarkable consequence, something which is so well known today. In the Newtonian theory, a particle at rest has no energy; the formula $E = (1/2)mv^2$ gives $E = 0$ when $v = 0$. But, in the formula Einstein obtained, even a particle at rest has energy called 'rest energy' and is given by the famous relation $E = mc^2$. For normal material particles, like a ball or a car, this is not of much consequence because this energy cannot be "tapped". You cannot make the particle lose some of its mass and convert part of its rest energy into something useful. But as we know today, this is precisely what takes place in nuclear reactions. If a nucleus A is split into two other nuclei B and C, it can happen that the total mass of B and C is slightly less than the mass of A. Similarly, one can also combine two nuclei X and Y to make a new nucleus Z with the mass of Z being slightly smaller than the sum of the masses of X and Y. In either case, the difference in the mass gets converted into energy according to the above formula. (The splitting of a nuclei into two is called 'fission' while combining two nuclei is called 'fusion'.) Nuclear reactors as well as atomic devices use these processes to generate enormous amount of energy. (We shall say more about this in Sec. 3.6.) In nature, this is precisely the source of energy which makes the Sun and all other stars shine. In Sun, the fusion of hydrogen nuclei to form Helium nuclei is the main source of energy. So if relativity was wrong, Sun will not shine!

More generally, relativity identifies energy with mass and treats them as two entities which can be converted to one another. This is a radical departure from Newtonian mechanics. Before relativity came along one viewed mass, which determines how a body responds to external forces, as quite distinct from energy of any form. Electromagnetic radiation, for example, is thought of as a pure form of energy but one never attempted to attribute an equivalent mass to the electromagnetic radiation. Relativity changed all this and opened up the possibility that matter can be converted into energy and vice-versa. As we will see in Chapter 4 this opened up a completely new way of looking at nature.

There is another feature which relativity introduces into the description of nature. In Newtonian mechanics, information can propagate instantaneously, that is with infinite speed. Suppose the structure of the Sun changes at some moment of time, thereby changing the gravitational force exerted by Sun on Earth. In Newtonian mechanics this change will be experienced at Earth instantaneously. This violates our result that no signal can propagate faster than light. Light takes about 8 minutes to come from Sun to Earth; so any change in the structure of Sun can be noticed at Earth only after at least 8 minutes. The original laws of physics in which signals could travel with infinite speed need to be modified because relativity forbids this.

What are these laws? To begin with, classical mechanics had the Newton's laws of motion and we have seen above how this can be modified in the wake of relativity. Essentially, one just needs to change the relation between momentum and velocity as well as that between the energy and velocity. With this modification, everything works out fine. Second, we also had the laws governing electricity and magnetism which were described by Maxwell's equations. One might have liked to check them out and modify them in the light of our new found knowledge. Incredibly enough, this was not needed! The Maxwell's equations were perfectly correct and were quite consistent with relativity. As you could have guessed, this meant that Maxwell's equations had some problem with the simple formulation of Newtonian physics before relativity came along. In fact, it is the study of these difficulties which lead to the theory of relativity rather than experimental results. As you might recall from the last chapter, Maxwell's equations describe the propagation of light as an electromagnetic phenomena. Since the propagation of light and its speed plays a crucial role in the formulation of relativity, it is gratifying that Maxwell's equations come out unscathed in the relativistic revolution.

What remains is the description of the gravitational force in terms of Newton's laws of gravity. Newton's theory tells you that the gravitational force propagates with infinite speed which is clearly in violation of relativity. (In contrast, it should be stressed that Maxwell's equations require electromagnetic force to propagate at the speed of light.) Moreover, the source of gravity in Newtonian theory is the mass of the particles and we just saw that mass and energy are equivalent in special relativity. If mass produces a gravitational field around it, so should all forms of energy.

These considerations suggest that we need to tweak the theory of gravity little bit to make it consistent with relativity. It turns out, however, that no simple tweaking will help and one requires a fairly drastic modification of our ideas of space and time. The resulting model is known as "General theory of Relativity" and incorporates the behaviour of gravity in an extraordinarily beautiful manner.

2.4   Principle of equivalence

To understand why combining special relativity with Newtonian gravity leads to another revolution, one has to review some peculiar features of Newtonian gravity.

The Newton's law of gravitation tells you that the gravitational force exerted by a body of mass M (say, the Sun) on another body of mass m (say, the Earth) is proportional to the product Mm and decreases as the inverse square of the distance between the bodies. What we want to concentrate on is the fact that the force is proportional to the product of the two masses. When *any* force F acts on a body of mass m, it produces an acceleration $a = F/m$ which is just Newton's second law of motion. Usually the force will not depend on the mass of the body on which it is acting. Hence the acceleration will be larger for lighter bodies and will be smaller for heavier bodies. But in the case of gravity, we just found that the force itself is proportional to m. Therefore, the gravitational acceleration is independent of the mass of the body.

This is actually a rather peculiar situation. If you drop two balls of masses 5 kg and 1 kg from the top of the Leaning Tower of Pisa (Galileo is often credited with doing it though he never really performed this experiment), then both will fall with the same acceleration $g = 9.8$ m s^{-2} and will hit the ground at the same time. This happens precisely because the gravitational force scales in proportion with the mass. The situation is shown schematically in Fig. 2.3(a). If you drop two balls of different masses inside a cabin standing on the surface of the Earth, the balls will hit the ground at the same time.

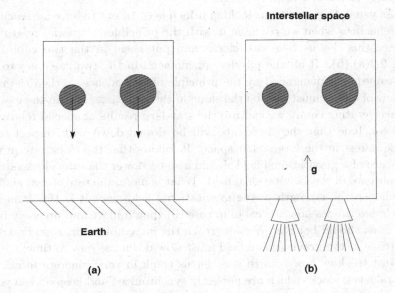

Fig. 2.3   The equivalence between gravity and accelerated motion. The left frame shows a cabin at rest on the surface of earth while the right frame shows the same cabin accelerating upwards in the interstellar space. Two balls of different mass, dropped inside the cabin will hit the ground simultaneously in both cases.

There is, however, a completely different physical context in which the same result can be obtained. This is shown in Fig. 2.3(b). Suppose you are inside a cabin which is moving in interstellar space away from all gravitating matter and is accelerating upwards with an acceleration $g = 9.8$ m s^{-2}. If you now drop two balls of different masses inside the cabin, they will not move because there is no force acting on them. However, the floor of the cabin will move up with the acceleration g and will hit the two balls simultaneously. It has to be simultaneous because the floor of the cabin knows nothing about the two balls or their masses. But when you are inside the cabin and moving with it, we will interpret the floor to be stationary with respect to you and the two balls as dropping down with the same acceleration. Having done similar things in Pisa, you will imagine that you are actually residing in a cabin located on the surface of the Earth. In summary, this experiment cannot distinguish the physical phenomena in an accelerated cabin from physical phenomena taking place in a gravitational field. This goes under the name *Principle of Equivalence* and is a unique property of the gravitational field.

As usual, this innocuous looking principle turns out to have far reaching consequences when we combine it with the principles of special relativity. To see this, let us take two clocks and put them in the two cabins in Fig. 2.3(a),(b). If all the physical phenomena in the two cases are to be the same — as demanded by the principle of equivalence — then the rate of flow of clocks must also be the same in the two cases. But in the case of an accelerating cabin, we can use the standard results of special relativity and conclude that the clock rate will be slowed down with respect to a clock at rest in the interstellar space. It follows that the clock rate in the presence of a gravitational field should also be slower than the clock rate in the absence of the gravitational field. What is more, the rate of flow should depend on the strength of the gravitational field since it is this strength which determines how the cabin in case (b) should move and precisely how the clock should behave. We thus reach the incredible conclusion that the existence of the gravitational field must slow down the flow of time!

Just to show how bizarre it is, let us think in very concrete terms. If you take two clocks which are perfectly synchronized and keep one on your table and the other on the ground, then the one on the ground will run slower compared to the one on the table. Special relativity told us that a *moving* clock will run slower compared to a stationary clock and that was bad enough. Now when we add the principle of equivalence to the fray, we end up concluding that even two clocks *at rest* with respect to each other will run at different rates if they are located at different gravitational fields.

Since we use light rays to measure distances, what affects the clock time also affects the measurement of distances. In the presence of the gravitational field, both the measurement of length and time becomes modified and varies from place to place. One can no longer talk about absolute time or absolute length without specifying what kind of background gravitational field one is encountering.

At this stage the mathematics becomes incredibly complicated and — unlike special relativity — it is not easy to develop the formalism in qualitative terms using only English and logic. Nevertheless, we will try to describe the key ideas which this formalism brings in without using any mathematics. The key new concept which now enters is the so called *curvature of space and time* and we will try to understand what it means.

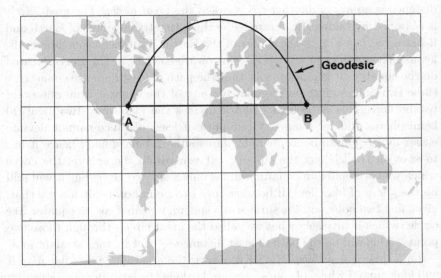

Fig. 2.4 The shortest path ('Geodesic') between two points on the globe which are located at the same latitude. The curved line is the shortest path and *not* the latitude connecting the two points A and B. The curved line is actually 'straight' but appears curved because it is drawn on the surface of a sphere which itself is curved.

The easiest way to understand curvature is to compare the surface of a sphere with the surface of a sheet of paper. Both are two dimensional surfaces in the sense that any point on either surfaces can be specified by giving two distinct numbers. For example, the longitude and latitude will uniquely specify a point on the surface of the globe which is a sphere. The intrinsic geometrical properties on these two surfaces, however, differ significantly. Take, for example, the notion of a straight line between two points on the globe which, for clarity of illustration, we will take to be at the same latitude. If you view the globe as a flat map hanging on the wall, then the straight line connecting two points with the same latitude is just the one moving along constant latitude (see Fig. 2.4). On the spherical globe, this will correspond to an arc of a circle drawn parallel to the equator with its centre on the axis of rotation of Earth. As we move away from the equator to higher and higher latitudes, the corresponding circles will have smaller and smaller radii. Incredibly enough, this is not a "straight line" on the surface of the globe; that is to say, this is not the path of least distance between the two points. The curve connecting the two points having least distance is an arc of another circle called the *great circle*. The great circle is obtained by taking the centre to be the geometrical centre of Earth and it has a radius equal to the radius of the Earth. To understand this result, let us first consider any two locations on the surface of the Earth located on the equator. It is now obvious that the path of least distance connecting these two points must be the (shorter) arc of the equator from one point to another. This is because equator divides the globe into two identical hemispheres and by symmetry our path of least distance cannot deviate either into northern hemisphere or into southern hemisphere; hence it has to stay on track along the equator. If you rotate the sphere, the curve which was originally an equator will become a slanted circle but it will still be the curve of least length between any two points on it. It follows that, given any two points on the surface of a sphere, we can draw an equator-like circle (which is precisely what we called the great circle) through those two points. This will be the path of least distance — that is, the "straight line" — on a spherical surface. But, if you draw the great circle in a flat map it will look curved while, of course, the circle along the latitude looks straight. This is the first jolt you get while dealing with curved surfaces in contrast to flat surfaces. The idea of straight line is often not intuitively obvious.

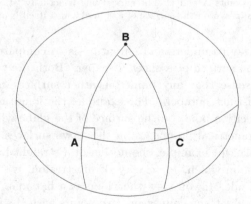

Fig. 2.5   Laws on geometry on a curved surface are different from those on a flat sheet of paper. Figure shows a triangle ABC drawn on the surface of earth using three 'straight' lines; two latitudes BA and BC where B is the north pole and the equator AC. Clearly, the sum of the three angles of the triangle is greater than 180 degrees.

The weirdness does not stop there; our simple ideas and theorems from geometry no longer hold in a curved surface. Figure 2.5 shows a triangle drawn on the surface of a globe. It is a triangle in the sense that three points A, B, C are chosen on the surface of the globe and and they are connected by paths of least distance which are the straight lines. (The BA and BC are two longitudes which are parts of great circles and AC is part of the equator, which is also a great circle.) It is clear from the figure that the angles A and C are 90 degrees and angle B could have some other value, say, 30 degrees. If you add up the three angles of the triangle ABC, you get a result which is larger than 180 degrees while for any triangle drawn on a plane surface it should always be 180 degrees. One usually says that when the space is curved we have to use the results of *non-Euclidean* geometry while the standard theorems you learnt in the school correspond to Euclidean geometry.

The above result also shows that the usual ideas which we use to measure distance between two points, based on the Pythagoras Theorem, will not work. (After all, if a triangle has two angles both of which are 90 degrees, you will have a tough time defining a hypotenuse!) In fact, the distortion of areas of continents — Africa hardly looks the second largest continent in a flat world map — is related to this feature. One needs to give specific rules to measure distance between any two points on a curved space and these rules depend on the kind of curvature the space has. Mathematically, this is done by giving a set of functions called *metric* using which one can work out all the properties of a given curved space.

So far, we talked about two dimensional spaces because they are easy to visualize. In relativity, an event is characterized by four numbers and hence is a point in a four dimensional spacetime. But a space of any dimension can be flat or curved. In special relativity, the four dimensional space, made of events, is assumed to be flat and one can use the laws of Euclidean geometry to describe measurement of length and time intervals. But as we saw before, the gravitational field affects the rate of flow of clocks and the measurement of space intervals. Specifying the distances and time intervals between different events in spacetime will be equivalent to thinking of the spacetime as curved. A simple way to state this result is to say that gravity makes spacetime curved. That might sound mysterious but all that it means is the following: In the presence of a gravitational field, the standard rules of Euclidean geometry cannot be used to measure distances and time intervals between different events. The new set of rules can be specified in terms of a set of mathematical functions (called the metric) and the gravitational field is encoded in the form of the metric of the spacetime.

This might sound quite far away from the real life in which the gravitational field is thought of as a force acting on a particle. The Earth, for example, goes around the Sun because, at every instant, the Sun is exerting a gravitational force on it. If the gravitational field of the Sun disappears suddenly, the Earth will shoot off in a straight line. How do we describe the same phenomena in the new language? Actually, it turns out to be quite simple and elegant. We first say that the Sun curves the spacetime around it; which is just another way of describing the effect of Sun's gravity on the rate of flow of clocks and measuring rods. We then say that the Earth is not acted upon by any force but merely moves in a "straight line" in the resulting curved space. As we saw in Fig. 2.4, a "straight line" in curved space might appear curved to us who are so accustomed to flat space. Just look at the line AB in Fig. 2.4 which is the path of least distance on the surface of the globe but doesn't quite appear to be straight. This is precisely what happens in the case of Earth moving around the Sun. It is quietly following a straight path — but in a curved space — and we think that it is a curved path (like an ellipse) in a space which we assume to be flat. In this approach, one can completely forget about the gravitational force acting on particles and instead think entirely in terms of curvature of spacetime. This is how Einstein described gravity.

Once again, the ultimate test of any such physical theory is verification by observation. In the case of gravity there are several situations in which Einstein's theory will lead to results which are different from the corresponding ones in the Newtonian theory. In each of the cases, Einstein's theory has been vindicated which makes people believe that this is the correct description of nature. Let me just describe a couple of these crucial tests of Einstein's theory.

The first result has to do with the actual path taken by a planet while going around the Sun. In Newtonian theory, this is essentially an ellipse under certain conditions. When you calculate the trajectory under the same conditions using Einstein's theory, you will find that it is not an ellipse. Instead, it is what is known as as precessing ellipse, shown in Fig. 2.6. The planet keeps orbiting the Sun almost in an elliptical path but the orientation of the ellipse itself slowly changes with time. (This is called *precession*.) There are several other effects in the solar system, in particular the gravitational field of other planets on a given planet, which will lead to the same phenomenon. In the case of Mercury (which is the closest planet to the Sun for which the influence of the Sun is strongest), people had calculated the rate of precession due to the effect of other planets. The theoretical value was off the observed value, when all the calculations

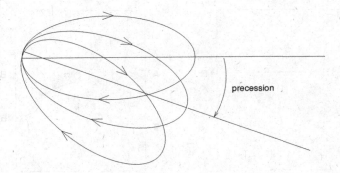

Fig. 2.6 The path of a planet around the sun according to general relativity, with scales grossly exaggerated for clarity. The orbits are (approximately) ellipses but the orientation of the ellipses changes over time with the major axis of the ellipse shifting slowly.

are done in Newtonian theory, by a small amount of 43.03 arc seconds per century. (Astronomical observations have been going on for a sufficiently long time for this disparity to have been noticed!) When you redo the calculation using Einstein's theory of gravity, one obtains an extra bit of precession which nicely brings the observations and theory into agreement.

The second result has to do with the behaviour of light in a gravitational field and it can again be illustrated nicely by using our analogy between an accelerated cabin and the gravitational field. Figure 2.7 shows the accelerated cabin with a source of light beam A on its left wall. Let us assume that a pulse of light is emitted from this source and travels towards the opposite wall. If the cabin was not accelerating, it will, of course, travel in a straight line and hit the opposite wall at the point B. But as the light pulse travels between A and B the cabin is accelerating upwards and would have moved a little further. As a result, the light beam will hit the opposite wall at a point B' little lower than B and its trajectory will be as shown in AB'. An observer inside the cabin will, of course, see that the light is travelling on a non-straight trajectory but she will also assume that the cabin is located on the surface of Earth which is exerting a gravitational force downwards. She will therefore conclude that the gravitational field can act on light beams and make its trajectory bend. More generally, light will deviate from a straight line trajectory when it passes close to a gravitating body.

When the amount of deviation is computed, one finds that Einstein's theory predicts a value which is precisely twice the result predicted by

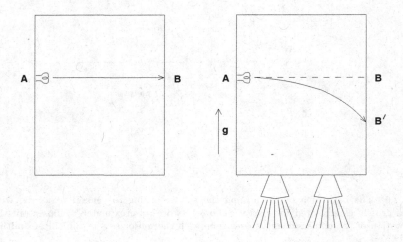

Fig. 2.7   The bending of light by gravity. The left frame shows a beam of light which travels horizontally in a straight line within a cabin at rest. The right frame shows the situation when the cabin is accelerated upwards. By the time the light reaches the other end, the cabin has moved by some amount making the light take a curved trajectory. Since the gravity acts in a manner similar to the accelerated frame, gravity should also exert a force on light beam, bending its trajectory.

Newtonian theory. For a light ray which is just grazing the surface of the Sun, the deflection due to the gravitational effect of the Sun is about 1.75 arc seconds. So the position of a distant star located just at the rim of the Sun will appear to have changed by this amount when viewed from Earth (see Fig. 2.8). Of course, you can't see a star which is just near the rim because of the bright glare from the Sun. But during a solar eclipse we are shielded from the Sun's rays and one can perform this observation. This is what was done in the year 1919 and the preliminary results were interpreted as vindicating Einstein's theory — making him famous overnight.[2]

2.5   Gravity and light: More amazing consequences

The fact that light bends while passing through the gravitational field of a massive body has far reaching implications. To understand that, we need to introduce the concept of a *light cone* centered on an event. Consider an

---

[2]Some general relativists are fond of saying that the deflection of light by the Sun is a very tiny effect. This is, of course, quite incorrect. Radio astronomers routinely work with the accuracy of milliarcsecond for source positions and any one of them will tell you that this effect is huge.

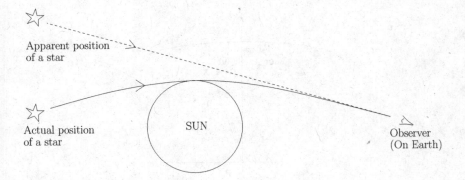

Fig. 2.8 The bending of light due to gravity changes the apparent position of celestial object as seen in the sky. For example, the bending of a ray of light coming from a distant star to Earth will make the star appear at an apparent position which is different from the actual position.

event P from which we emit light in all directions at time $t = 0$. After a lapse of some time t, the light front would have reached a spherical surface of radius $r = ct$. This is illustrated in Fig. 2.9 in which we have introduced time as a coordinate axis in the vertical direction and two spatial dimensions x, y along the horizontal plane. (In real life, we need to have three spatial dimensions but, of course, we cannot draw such a figure.) You can think of the conical surface in Fig. 2.9 as being generated by the location of the light front as it propagates.

Let us suppose you also threw a stone in some direction at $t = 0$ just when the light pulse was emitted. Since the stone (or, for that matter, any material body or signal) can only travel at speeds less than the speed of light, the stone can never catch up with the light front or pierce it. The trajectory of such a stone, PX will be a line completely inside the light cone in Fig. 2.9. If the stone is moving with uniform speed, the trajectory will be a straight line. But even if it travels with variable speed it can never overtake the light front and it will always be confined by the light cone.

Thus the light cone drawn at any given event P divides the space into two regions. All the points inside the light cone can receive signals from P. Those which are outside the light cone cannot receive signals from the event P, making the light cone act as a *one-way membrane* as far as signals are concerned.

All this is true in the absence of gravity. But when there is a gravitational field present in space, the light rays bend and no longer travel along

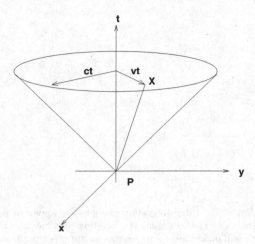

Fig. 2.9   The concept of a light cone. This is a 'spacetime diagram' with the vertical axis denoting time and the two horizontal axis representing two spatial directions; the third spatial direction is not shown. A beam of light emitted at P will travel along the conical surface and at time t will be located on a circle of radius ct. Any other material body will travel slower than the light. Another body thrown from P with speed v will be at the location X at time t, with its trajectory remaining always inside the light cone.

straight lines. This means the light 'cones' are no longer cones and can take a rather distorted shape. Near a very massive body, for example, the light cones will be tilted enormously towards the body and the light, if it has to escape from the body, has to travel within a small opening. Things get worse in the case of very strong gravitating objects called *black holes*. At the surface of a black hole, all the light cones are tilted inwards so that no material body can escape from the region inside the black hole to the outside. One possible way of understanding this result qualitatively is to again go back to our example of a pulse of radiation emitted from the origin at $t = 0$. In the absence of gravity, the light front travels outward with the speed of light and at a later time t will be located on the surface of a sphere of radius $r = ct$. Suppose now that there is a source of very strong gravity near the origin which is pulling the light front backwards. In the case of a black hole the outward motion of the light front is precisely balanced by the inward pull of gravity so that the light front remains at a given radius $r = R_s$ (called *Schwarzschild radius*) at all times. The original idea that the stone thrown from the origin cannot catch up with the light front continues to be true; in the current situation any material body or signal from the region $r < R_s$ can never go outside to the region $r > R_s$. The more massive a body is, the larger will be the region over which it can exert this influence. The formula for the Schwarzschild radius is $R_s = (2GM/c^2)$ where M is the mass of the body. When a body of mass M contracts to a size smaller than its Schwarzschild radius, it becomes a black hole.

There is a more graphical way of seeing this result which is originally due to Bill Unruh. Consider a river which flows, say from left to right, and ends up as a waterfall at some stage. We know that the speed with which the water flows in the river will increase as we approach the water fall. So the speed has some small value as we move further and further upstream and it increases and reaches a maximum value downstream, just before the water vertically plunges down as a waterfall. Let us assume there are different kinds of fish which live in this river, each of which can swim with different maximum speeds. We will call the best swimmers amongst them as "light fish" which can swim with a speed v_{max}, say. No other fish can swim faster than our light fish.

Consider now what happens when the light fish tries to swim upstream from some location in the river. If the speed with which water in the river is flowing is less than v_{max}, the light fish will have no problem swimming against the current and reaching wherever it wants to. But, as we noticed before, the speed of the river flow keeps increasing downstream and at some point — let us call it H — the river will flow with a speed v_{max}. To the right of H river will flow with a speed larger than v_{max}. If a light fish ventures into a region to the right of H, it will find itself in a dangerous situation. Since it can only travel with a maximum speed v_{max}, while the speed of the river downstream is larger at the location of the fish, it will never be able to successfully swim upstream. The river will drag it down further and further downstream and it will eventually fall with the waterfall. The light fish which is precisely at H — and trying to swim to the left — will find that it moving neither to the left nor to the right. It is trying desperately to swim upstream but its speed is exactly balanced by the speed with which the water is dragging it downstream. As a result, it will stay put at H.

The analogy with the black hole should be clear to you by now. The light fish is analogous to photons which travel with the speed of light and all other fish corresponds to material bodies which can only travel with speeds less than that of light. The region to the right of H is the interior of the black hole while the region to the left is the exterior. Material bodies and light can easily move from left to right and fall into the black hole just as the fish can swim downstream. On the other hand, no fish including light fish can escape from a region to the right of H upstream just as no material body or even light can move from the inside of the black hole to the outside.

The simplest context in which black holes occur in astronomy is at the end stages of the evolution of massive stars. Stars, for most of their lifetime (which is typically several billion years), shine by converting mass into energy through nuclear fusion. The Sun, for example, is acting as a giant nuclear reactor fusing hydrogen into helium in its core and releasing the energy. Eventually, this process exhausts the available source of nuclear fuel in the stars and the gravitational force will start making the stellar material contract. This collapse of the star under its own weight can be resisted most of the time by internal pressure and the star reaches its old age without becoming anything more exotic. However, if the mass of the star exceeds a critical value (first worked out by S. Chandrasekhar, which won him the Nobel prize) then gravity wins over the internal pressure and the star continues to collapse. Soon the gravitational pull on the surface of the star is so high that nothing, not even electromagnetic radiation, can escape from it. Typically, a star like the Sun has to contract to a radius of a few kilometers for it to become a black hole. The density of such an object will be enormous and will be more than the density of atomic nuclei. It is possible for black holes to form in more extraordinary scenarios but this is the most common situation.

You might be wondering, "But all this sounds purely hypothetical ! If no signal can come to us from a black hole, how will we know it exists?" Fortunately, the existence of black holes can be determined through indirect means. This is, of course, more difficult and prone to uncertainties than direct observations, but astronomers have succeeded in gathering fairly strong evidence for the existence of black holes in several cases. One of the signatures is the following: Most of the stars occur in what are known as binaries, which essentially consists of a pair of stars orbiting around a common center of mass, under their mutual gravitational attraction. When one of these stars evolves more quickly than the other and forms a black hole, it will exert a very strong influence on the companion star which is still an ordinary star. Material from the companion star will be pulled out due to the gravitational force of the black hole and will fall into the black hole, as the second star orbits around the black hole which itself will be invisible. By studying the orbital properties of the companion star, which of course emits radiation, one can infer the details regarding its invisible partner. In particular, if the mass of the invisible partner is significantly higher than the Chandrasekhar limit, then it has no choice but to be a black hole.

Black holes are probably one of the few astronomical objects which have caught the fancy of both the public and the science fiction writers. As a result, there is lot of "hype" surrounding the black holes ("space and time comes to an end in a black hole", "one can reach another universe through a black hole" ....) which makes them exciting but most of the more esoteric claims are not usually backed by concrete calculations. Occasionally, this also leads to misleading statements. For example, there are several black holes which exist in our own galaxy and, in fact, there is strong evidence to suggest that there is huge super massive black hole at the center of our galaxy. None of this has made space and time to come to an end for all of us! Such statements originate from the fact that, at present, we do not have a well defined theory which describes what happens to the matter which collapses to form a black hole once it is inside the black hole. This is because, as the matter compresses to smaller sizes, quantum mechanical effects play a greater role. To understand the end stage of matter which collapses, one needs a theory which combines the principles of general relativity and quantum mechanics which is currently lacking (see Chapter 6). This has led to people developing different kinds of ad-hoc models including very exotic ones in which the matter that collapses to form a black hole reappears in another universe. At present these are very speculative models and one cannot ascertain their validity.

2.6 Gravity and the universe

Changing the description of gravity has another incredible consequence: It changes our view of the cosmos. This has been a peculiar feature of gravity known, for example, even to Newton, except that with gravity described as the curvature of spacetime, *à la* Einstein, the description acquires new twists and turns.

This connection between gravity and cosmos arises from the fact that gravity is always attractive and cannot be shielded. This is one key difference between, say, electromagnetism and gravity. As you will recall, the electromagnetic force between two like charges is repulsive while that between unlike charges is attractive. So if you have a neutral object like a hydrogen atom made of positive and negative charged particles, it exerts very little electromagnetic force on the outside world. Essentially, the effect of positive charge and the negative charge cancel each other. In the case of gravity, on the other hand, there is no such thing as positive mass and negative mass. All matter gravitates with an attractive force. So when you consider larger and larger regions of the universe containing more and more matter, the gravitational effect keeps increasing and exerts a strong influence on the way matter behaves. In short, the very large scale dynamics of the universe is entirely governed by the gravitational force. Newton, for example, knew this and tried to construct models of the universe using his theory of gravity but it was not of much practical consequence since neither the theory nor observations were mature enough to comprehend the cosmos in those days. With Einstein's theory, it was different.

To understand the broad implications, let us begin by asking how the universe looks at very large scales. The simplest assumption would be to imagine that the universe is infinite in all directions and filled with matter uniformly. Observations suggest that this is actually a pretty good approximation at large enough scale. The situation is similar to our experience with regards a piece of solid, say, a chalk piece. The chalk piece appears quite smooth and one can attribute to it a constant density all throughout. But, of course, we know that any such solid is made out of atoms and molecules and, in fact, most of the space inside an atom is empty. The density of the nucleus on the other hand is enormously large compared to the average density of the chalk piece and one can no longer think of the chalk piece as a smooth object at molecular scales. In the case of the universe, the situation is exactly similar. Matter in the universe is made of stars, galaxies, clusters of galaxies etc. (we will say more about these in Chapter 5) just as a chalk piece is made of individual atoms. But when you consider a sufficiently large region of the universe containing several billion galaxies, the universe appears to be smoothly filled with matter. This is just like a chalk piece appearing smooth when we look at it from scales much bigger than the size of individual atoms. What we are interested in is the bulk dynamics of such a smooth universe containing a large number of galaxies, say.

This will be governed by the gravity generated by the smooth distribution of matter in the universe. In Einstein's theory, of course, there is no such thing as gravitational field or gravitational force; instead, the gravity is described by the curvature of the space. So we need to ask how exactly the smooth distribution of matter curves the resulting spacetime. There is a precise mathematical way of doing this within the frame work of Einstein's theory and the answer is mind boggling. The solutions to Einstein's equations in this particular context suggest that such a universe cannot be static; it has to be continuously expanding! In physical terms, such an expansion implies that if you are sitting on any one galaxy, you should see all other galaxies at sufficiently large distances to be flying away from you. This model of the universe is what is called the big bang model and has formed the cornerstone of contemporary cosmology. We will describe it in greater detail in Chapter 5 but it is probably worthwhile to clarify a few conceptual points at this stage itself.

The idea of an expanding universe can be pretty confusing when you first encounter it. The first impression one gets — and this is strengthened by the term "big bang" — is that since all the distant galaxies are flying away from us, we are at the center of the universe and the big bang explosion must have occurred right here. This idea is totally incorrect! Let us consider the two aspects of the question separately. To begin with, the fact that all the distant galaxies are moving away from us does not necessarily mean that we are at the center of the universe. In fact, if we live on another distant galaxy X, we will again see the same phenomena, viz. that all the galaxies are again moving away from X. To understand this, it is helpful to consider a simpler example. Imagine a very long stretch of road with cars moving in one direction. Assume that, at some instant of time, the distance between any two consecutive cars is 100 meters and that you are sitting in some chosen car, say, A (see Fig. 2.10). Let the first car (B) in front of you move at 10 km hr^{-1} with respect to you, the second one in front (C) at 20 km hr^{-1}, third (D) at 30 km hr^{-1} etc. It will appear to you as though all the cars are moving away from you with the speeds proportional to the distances from you. But consider a lady sitting in the car (B) just in front of you. How will she perceive the car just in front of her? Since she is moving at 10 km hr^{-1} with respect to you and the second car (C) in front of you is moving at 20 km hr^{-1} with respect to you, the driver of the car (B) will think that the car (C) is moving only at 10 km hr^{-1} with respect to her. By the same reasoning she will conclude that car D in front of her is moving at 20 km hr^{-1} etc. In other words, she will come to the same conclusion regarding the motion of vehicles as you did! Clearly, there is nothing special about your position. All the cars in this chain enjoy equal status.

The motion of galaxies in the universe is quite similar. Every observer, irrespective of at which galaxy she is located in, will see all the galaxies moving away from her with a speed progressively increasing with distance. This curious motion of the galaxies is what really constitutes the "expansion

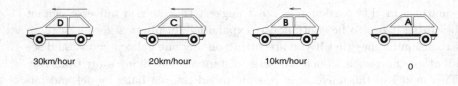

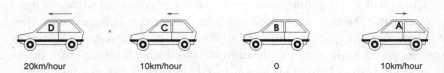

Fig. 2.10   A sequence of cars moving along a long stretch of road with the speeds shown in the figure. The top part of the figure shows the situation as viewed from the car A. The driver of A sees the three other cars moving with speeds which increase in direct proportion to their distance. The driver of this car might think that her position is "special" and that all the cars in front are moving away from her. But the driver of car B will see the car C to move at 10 km hr^{-1}; this is because car B is moving at 10 km hr^{-1} with respect to car A and car C is moving at 20 km hr^{-1} with respect to car A. Clearly, the driver of car B will also think that all the cars are moving away from her car.

of the universe". After all, the word "universe" is merely a convenient terminology for describing the matter which we see around us. It is the movement of the parts of the universe which is indicated when one talks of the expansion of the universe.

Since no galaxy in the universe is special, there is no "center" to the universe. In the above example, the road is assumed to be infinitely long and any car can be thought of as being in the center — which is the same as saying no car is special and that there is no real center. The same idea applies in two dimensions or three dimensions. Think of a rubber sheet with vertical and horizontal grid lines drawn on its surface. If the rubber sheet is stretched in the vertical and horizontal directions, the grid will stretch with the rubber sheet. If the sheet is infinite in both the directions, then again there is no point on the rubber sheet which can be thought of as the "center". All the points are equivalent. Hence, we should not think of big bang as some kind of a fire cracker explosion which took place *at a given point in space*. The entire space collapses to a point at big bang rather than space existing independently of matter and the matter collapsing to a point within that space.

The second popular misconception about an expanding universe is more subtle. You might have read in popular articles of scientists talking about the expanding universe using the analogy of a balloon. As air is blown into the balloon and it expands, any point on its surface can be thought of as a center with all other points moving away from the chosen point. This nicely illustrates the explanation in the previous para but raises another difficulty; you may wonder: "The balloon is expanding into the three dimensional space. What is the universe expanding into ? Is it expanding into higher dimensional space ?"

The balloon analogy is quite nice but — like all analogies — should be used only for the purpose it was intended. Suppose there are two dimensional creatures living on the surface of the balloon who are incapable of perceiving the third dimension. If these creatures measure the distance between any two points A and B on the surface of the balloon at two different times, they will find that this distance has increased. The cosmologists among them will interpret it as expansion without ever having to use the third dimension. Similarly, the manner in which we have described the expansion of the universe makes the question: "what is the universe expanding into ?" irrelevant. All we know from the observations is that distant galaxies are moving away from us with speeds increasing in direct proportion with the distance. Surely, in an infinite universe such a motion can take place without leading to any contradiction. If we understand the term "expansion of the universe" as representing "motion of distant galaxies away from us" then such confusions will not arise.

You might say at this stage: "But wait a sec! How can the two dimensional creatures perceive that the distance between A and B is increasing? They will have to use a scale or tape measure which itself will expand! So they will never know that the balloon is expanding!" But, no! In the balloon analogy we are assuming that the size of the creatures or their measuring instruments do not expand when the balloon expands, thereby allowing them to make the measurements. In the real universe this question translates into a valid doubt: If everything in the universe is expanding (galaxies, stars, earth, meter scales etc.) then how will we ever know that there is an expansion ? This is simply incorrect. Everything in the universe is *not* expanding. All that is happening is an increase in the distances between galaxies which are sufficiently far away from each other. The expanding model for the universe is a solution to the equations in Einstein's theory for gravity when one assumes that the matter is distributed *uniformly*. This is a good assumption at very large scales in the universe just as thinking of a piece of chalk as a continuous solid with uniform density is a good assumption at scales much bigger than the atomic scales. As we probe the chalk piece at smaller and smaller scales, we will eventually have to recognize the fact that it is made of atoms which themselves are made of electrons, protons and neutrons etc. One cannot attribute properties of the chalk (like its elastic strength) to the individual atoms which makes up the chalk. Similarly, the expansion of the universe is a property of the universe in the bulk, when the matter is treated as distributed uniformly and the smaller scale structures like galaxies — which are analogous to atoms in a piece of chalk — are ignored. As one proceeds to smaller scales, the existence of these structures need to be taken into account. In fact, the closest big galaxy to Milky Way, called "Andromeda" , is actually moving towards Milky Way rather than moving away from us. This is because the individual gravitational force between Andromeda and Milky Way dominates over the cosmic expansion. The description at small scales needs to take into account the gravitational effect of individual galaxies and we cannot use the results obtained for a uniform distribution of matter. The stars or the Earth or the meter scales which we use for measurements are *not* expanding and thus can be used to determine the expansion of the universe.

Finally, we conclude this section by narrating the story of what one should consider to be The Most Glorious Missed Opportunity of Civilization. We mentioned earlier that when you write down and solve Einstein's equation for a uniform distribution of matter in the universe, the solution will describe a universe which is expanding. Einstein, obviously, went through this exercise and had the chance to predict — based on pure mathematics and logical thought — that our universe must be expanding. (There was no observation indicating the expansion of the universe at that time.) Unfortunately, even Einstein faltered at this stage and thought this a little bit too radical. So he tweaked his equations by adding an extra term — called the cosmological constant — in order to stop the universe from expanding and to obtain a static solution. But alas, he made two mistakes in the process. First, purely mathematically, the extra term was not sufficient to make the universe static; other scientists, especially de Sitter, soon showed that even with the extra term one will have an expanding solution. Second, and more important, observations soon showed that universe is indeed expanding. It would have been wonderful to *predict* this feature rather than explain it after the observations.

Einstein did realize that he missed a great opportunity and called his tweaking the equation as the biggest blunder in his life. The final twist to the tale is that, in the last decade or so observations have suggested that Einstein, after all, could have been actually correct. We now believe that the cosmological constant introduced by Einstein and abandoned later on is indeed nonzero and poses a fascinating theoretical puzzle — which we shall take up in Chapter 5.

2.7   The odd man out

The key message you should get from all these is that the innocuous looking gravitational force is actually quite bizarre and is different from every other force (like, for example, electromagnetism) known to mankind. Since this peculiar behaviour of gravity remains one of the greatest challenges faced by theoretical physics today, it is worth summarizing some of its ingredients.

The first point is that you cannot construct a theory for gravity which is consistent with special relativity without giving up our standard notions of space and time. This arises from two facts, both experimentally established with great level of accuracy. First, the flow of time is relative and not absolute and moving clocks slow down. Second, one cannot distinguish the effects of gravity from the effects of acceleration in a local region. Since acceleration involves motion which affects the rate of clocks, one ends up concluding that flow of time is affected by the gravitational field. Because the measurement of length intervals is intimately connected with measurements of time intervals in relativity, we need to revise our notions of both space and time in the presence of gravity. As we described in Sec. 2.4, this revision leads to the conclusion that space and time are curved and it is this curvature which we mistake for gravitational force.

The second point is that gravity affects the propagation of light signals by bending the light rays. Strong gravitational fields can bend the light rays so much that they can prevent its escape from regions of spacetime. Because nothing can travel faster than light, gravity can thereby effectively shield a region from communicating with the rest of the universe. This leads to the notion of entities like black holes and is a key feature of Einstein's theory of relativity.

Finally, gravity cannot be shielded. So as we look at larger and larger regions of the universe, the gravitational influence of matter keeps on increasing and — at the largest scales — gravity emerges as the most important force. This demands that any model for the universe needs to be developed using the theory of gravity which we have at hand. When you do it with Einstein's theory, you end up with a model of the universe which is continuously expanding. Obviously the universe would have been smaller in the past and at some stage it would become a point and all kinds of bad things will happen. This poses yet another theoretical challenge essentially triggered by gravity.

So tackling one of the two clouds present in the horizon of physics in the late 19th century had led to all the above consequences. We shall next describe what the other cloud did.

Chapter 3

Madness of the micro-world

3.1 Waves

We will now take up the second major revolution of the 20th century, viz.,
the development of quantum theory. By and large, quantum theory refers
to the laws of physics which govern the behaviour of matter at small scales:
molecules, atoms, nuclei etc. Since all matter is made of these subunits,
one could take the point of view that the real world is completely quantum
mechanical and the classical physics (as opposed to quantum physics) is
just an approximation. The laws of quantum mechanics, as we will see,
are pretty weird and is nothing like what we are accustomed to in every
day physics. Nevertheless they govern not only the subatomic physics but
just about all of solid state physics and chemistry which forms the basis of
modern technology.

The key new feature which arises in quantum theory is the manifestation of wave nature of — what we have conventionally thought of —
as particles. And, complementing this, is the particle nature of what one
would have conventionally called a wave! For example, an electron, which
you may think is a particle, will behave like a wave under certain circumstances. Similarly, an electromagnetic wave will behave like a system made
of particles (called photons) in some contexts. Quantum theory allows us to
describe these two concepts at one go in a unified language thereby allowing
a better understanding of all phenomena.

Since we need to understand such a duality, it is better to begin by re-
calling some elementary properties of conventional wave phenomena which
we described in Chapter 1. We saw that one can associate with any wave
four different physical quantities: frequency, wavelength, speed of propagation and amplitude (see Fig. 1.11). We described them in Chapter 1 for

the electromagnetic wave but our discussion is applicable to any such wave. While all the four attributes of a wave (wavelength, frequency, speed and amplitude) will be of use in different contexts, it is the wavelength which is the most fundamental characteristic of a given wave. Among other things it is the wavelength that determines whether one can really observe the wave properties clearly.

To understand this, let us consider an example familiar to all of us. Suppose you are standing outside a building near one of its corners and two other people are coming around the corner from the other side of the building. If they are talking to each other, you will be able to hear them even though you will not be able to see them until they turn around the corner of the building. We know that both sound and light are transmitted as waves. The light waves from the two persons are not reaching you and you are not being able to see them. On the other hand, the sound waves generated during their conversation do seem to reach you, travelling around the bend. How is this possible?

The reason essentially has to do with the wavelengths of these two waves. Sound waves usually have wavelengths in the range of few meters which is comparable to every day dimensions. In the above example, the width of the passage way around the building will be comparable to the wavelength of the sound wave produced in the conversation. When this happens the waves have an ability to spread by a process called *diffraction* and will no longer travel in a straight line. This is why one could receive the sound wave from people talking from the other side of the building block. The sound waves diffract and bend around the corner.

The situation is quite different if the wavelength of the wave is very small compared to other dimensions involved in the problem. In this case, the waves oscillate far too rapidly in any given region of space and cannot diffract effectively. In such a context, waves essentially lose their "wave nature" and travel in a straight line almost like a particle. In fact, this is what happens to light most of the time. Light waves have extremely tiny wavelength varying between 300 to 600 nanometers where 1 nanometer $= 10^{-9}$ meter (i.e., one billionth of a meter). This is very tiny compared to normal everyday dimensions because of which the oscillations of the light wave is far too rapid to manifest itself. So the light waves from the people around the corner cannot diffract and reach you. That is why you can't see them but can hear them. This idea will turn out to be of fundamental importance in our description of quantum theory and hence worth remembering.

The concept of diffraction which we mentioned above is an inherently wave phenomenon. A closely related property of the waves is to interfere with each other. To understand this feature, consider two waves of the same wavelength which are present simultaneously in a given location in space. If the crests and troughs of the waves coincide with each other, then the waves will enhance each other and the amplitude will be doubled. (We say that the waves are "in phase".) On the other hand, if the trough of one of the waves coincide with the crest of the second wave, then they will cancel each other and there will be no net effect at that point. (The waves are "out of phase".) If we denote the crest by $+1$ unit of amplitude and the trough by -1 unit of amplitude, we can describe the two extreme cases as corresponding to $+1 + 1 = +2$ or $+1 - 1 = 0$. In general, adding up the two waves of unit amplitude you can get a result which varies anywhere from 0 to 2 units. When the waves annul the effect of each other, one calls it destructive interference; when they add up to produce twice the effect, one calls it as constructive interference, ignoring the oxymoron. The ability to interfere with each other is a distinctly wave phenomena. You can never add some tennis balls to more tennis balls making all the balls disappear. You can, however, add a wave to another wave and destroy both of them completely.

It is actually possible to observe the interference of light waves in several physical contexts — the beautiful colours in a thin oil film you see on the road are actually due to interference — and we will consider one particular case which will turn out to be quite instructive. The schematic description of the experimental set up is shown in Fig. 3.1. Light of a given wavelength λ is emitted by a source S and hits a screen AB which has two tiny holes X and Y. The sizes of these holes as well as the separation between them should be tiny and comparable to the wavelength of light. On the right side of the screen AB, most of the light is blocked and one can essentially think of X and Y as new sources of light waves. These sources illuminate another screen CD as shown in the figure. Consider now the intensity of light wave at any given point P on the screen CD. One receives the wave fronts from both X and Y at this point P. However, the distances PX and PY are in general different. As a result, the light wave would have undergone different number of oscillations travelling along XP compared to travelling along YP. For the sake of definiteness, suppose we choose a point P such that the distance XP is equal to ten wavelengths while the distance YP is equal to eleven wavelengths. (These are chosen just for illustration and are not realistic for laboratory experimentation; what really matters is the difference of the lengths XP and YP.) The wave from X would have

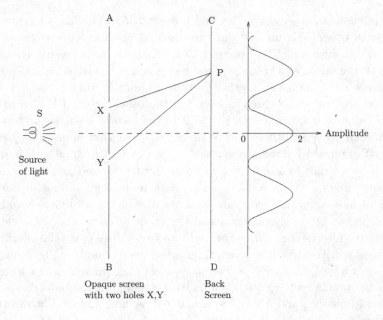

Fig. 3.1   Interference of light in a two slit experiment. Light from the source S passes through two slits X and Y and reaches a screen CD. The pattern of light intensity along the screen is shown as a graph on the right side of the screen. At points like P, where the waves from X and Y interfere constructively, the amplitude adds up and we get maximum intensity. At other points in between, the crest of the wave from one of the slits coincides with the trough of the wave from the other slit so that the sum total of the amplitude vanishes. At these points, there will be no intensity, showing that the addition of two light waves can lead to darkness.

completed ten oscillations on reaching P while the wave from Y would have completed eleven oscillations. But after an integer number of oscillations, the waves will have the same amplitude as before and hence the two waves will add in phase at P enhancing each other. As a result, the amplitude of the wave at P will be twice as large as the amplitude of the single wave. In contrast, let us consider another point P such that XP is, say, 13 wavelengths long while YP is 13.5 wavelengths long. Now the wave coming from Y will be half an oscillation ahead or behind the wave coming from X. So clearly, if the wave from X has its crest at P, the wave from Y will have its trough at P nicely cancelling out the effects. As a result, there will be zero electromagnetic wave intensity at this point and there will be no light!

As we move from the center of the screen CD either up or down, at each point the waves from X and Y will add up differently. In some places they will enhance each other leading to twice the amplitude and in some places they will destroy each other leading to complete darkness. Here is a remarkable result of light adding to light thereby producing darkness which — as we said before — is the key feature of wave phenomena. Such interference bands, as they are called, can be easily observed in the laboratory and was historically quite important in establishing the wave nature of light. The curve drawn at the right end of the figure Fig. 3.1 symbolically represents the intensity of light that you will find at any given point P. The wiggles show that there are regions where the waves enhance each other and there are regions at which they destroy each other.

3.2   Electron: Wave or particle?

We are now in a position to describe the central mystery of the microworld. This is probably best done using a thought experiment described, for example, by Feynman. To perform this thought experiment, we will set up the apparatus (as in Fig. 3.2) in a manner similar to what we did in the previous section. The only difference is that the source now does not emit light but a stream of particles, say, tiny steel balls. The two holes in the screen are bigger than the size of the balls and the third screen is equipped with some apparatus to capture the steel balls which hit it at any given point.

What do you think will happen in this experiment if we first close one of the holes, say X, and keep only Y open? Most of the steel balls will not get through the hole Y in the first screen and will just bounce off. There will be a small number of balls which will get through the hole Y and will land somewhere around Y' which is along the straight line SY. There could be few balls at other points on the screen, reaching due to scattering off the edges of the hole Y. Overall, the curve drawn at the right end of the Fig. 3.2 graphically illustrates the distribution of steel balls in this case with maximum number of balls being found at Y'.

Suppose we now close the hole Y and keep hole X open. Now some balls will travel along SX and hit the screen at X'. Of course, there could be a few more which scatters off, say, the edges of the holes etc. and hit the screen around the points X' thereby again smearing the distribution little bit. The curve Fig. 3.3 at the right shows this situation.

Let us now keep both the holes X and Y open and ask what happens? It will be pretty easy for you to reason along the following lines: "Take any

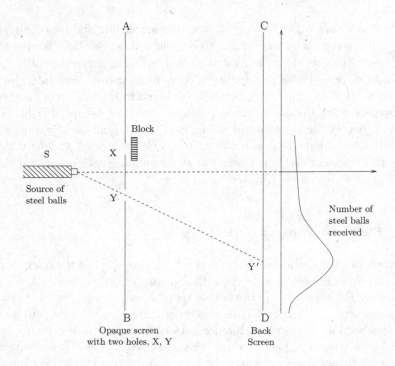

Fig. 3.2   A thought experiment using two slits and a source of small steel balls. The steel balls from source S passes through the slit Y and hits the screen CD where they are collected. The number of steel balls received at different locations on the screen CD is shown graphically just behind the screen. Most of the steel balls hit the point which is in the straight line SY while a few could get scattered at the slit producing the "tail" of the distribution.

given steel ball which is thrown out by the source. Three things can happen to it. It can just bounce off the first screen and not reach the second one at all; we are not concerned with these steel balls. It can go through hole X and form a pattern like in Fig. 3.2. Or, it can go through the hole Y and form a pattern like in Fig. 3.3. At any given point we will now have some steel balls coming through X and some coming through Y so the total number at any given point is just the sum of these two. The resulting distribution of balls on the screen will look somewhat like the pattern in Fig. 3.4." Obviously the number of balls which reaches any given point can now only go up compared to the case in which only one of the holes were open. If you do this experiment with real steel balls of, say, 2 mm in diameter, this is indeed roughly what you will see.

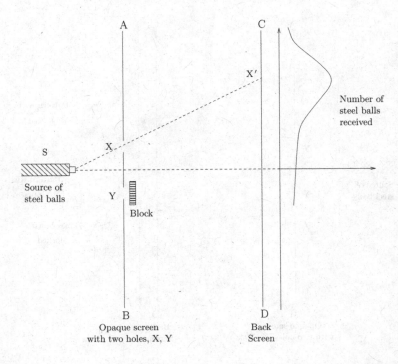

Fig. 3.3 The same thought experiment as in the previous figure but now performed with the slit X open and slit Y closed.

What happens if we now make the steel balls tinier and tinier or — better still — repeat the experiment with, say, electrons using an electron gun at S? When one of the holes is open and the other is closed, the electrons behave almost like steel balls and you get the patterns like in Figs. 3.2 and 3.3. But when you keep both the holes open something bizarre happens. You don't get a pattern which you got with steel balls but instead you get a pattern similar to that in the case of light like in Fig. 3.1. With two holes open, electrons behave like waves and interfere with each other!

The key feature to note is that there are points on the screen at which you get *less* number of electrons with both holes open than with just one hole open. Think about it. This is completely weird. If just one hole lets in certain number of electrons to reach a point on the screen, how can providing an alternate route for electrons to reach the point (through the second hole) *decrease* the total number reaching there?! You would have thought that a given electron either goes through hole X or through hole Y.

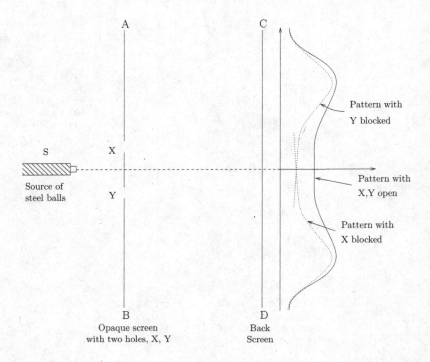

Fig. 3.4   The thought experiment with steel balls and the two slits X and Y. Now both the slits X and Y are kept open and the number of steel balls received on the screen has the pattern as shown. This is essentially obtained by adding the number of steel balls arriving at any given location on the screen through either of the slits individually. It is clear that the number of steel balls reaching any given point on the screen, with both slits open, will be larger than the corresponding number when only one of the two slits is open.

In that case, the total number of electrons which reaches any given point on the screen is simply the sum of the number of electrons that reach through hole X and that which reach through hole Y. With two holes open, the number can only increase with respect to the situation with one hole open.

People have pondered over this "mystery" and have tried all kinds of mechanisms to explain it. But it is impossible to understand it using the concept of particle we are accustomed to in everyday life based on steel balls. The only way out is to say that our assumption "a given electron goes either through hole X or through hole Y" is incorrect. One has to instead say that a given electron has some chance to go through hole X and some chance to go through hole Y. With these chances, one has to associate a wave and let these two waves interfere exactly as we did in the case of light. Then one will get the interference pattern for the electrons just as in the case of light waves.

So the new ingredient is the association of a wave with the propagation of an electron from one point to another. We need to constantly add these waves in order to figure out how exactly a given physical situation will develop. Figure 3.5 shows this idea in a more abstract context. Suppose an electron was at the point 1 at some time t_1 and was at point 2 at time t_2. If an electron were to behave like a steel ball, one would have associated a particular trajectory (or path) by which it went from point 1 to point 2. Our experiment has taught us that we cannot do this. We cannot associate a definite trajectory with an electron. It is allowed to take all possible trajectories between these two points as shown in Fig. 3.5. With each of these trajectories, we will associate a probability wave (the technical term is a probability amplitude) and to find out how the electron went from point 1 to point 2 we need to add up all these amplitudes in a particular way. These amplitudes can, of course, interfere with each other cancelling and enhancing the results in different contexts. What we finally observe is the result of this sum over all possible trajectories.

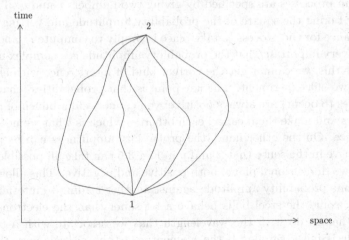

Fig. 3.5   In quantum mechanics, a particle does not follow a specific trajectory in moving from event 1 to event 2. Instead, we need to attribute some chance for it to take any given path connecting the two events in terms of a wave amplitude. The total amplitude that the particle will go from event 1 to event 2 can be computed by summing up the wave amplitudes for each of the individual paths.

In the above description we talked about the "chance" and "probability wave" for the electron to go from one point to another and this idea requires some elaboration. The actual rule in quantum theory governing this particular context is the following. One can associate — what is called — a probability amplitude, $\psi(2,1)$, for the electron to take a particular path to go from an event 1 to event 2. We are expected to sum the probability amplitude for all the paths connecting 1 and 2 and obtain a total probability amplitude, say, $K(2,1)$. The actual probability for this process to take place is given by the square of the probability amplitude K. The only complication is that the amplitudes like ψ, K etc. are in general what mathematicians call a complex number. [A complex number z can be expressed in the form $z = x + iy$ where x and y are ordinary numbers and the symbol i stands for the imaginary quantity $i = \sqrt{-1}$. Fortunately, we won't need this to understand most of what follows.] You can think of a complex number as something similar to a two dimensional vector shown in Fig. 3.6. Each complex number corresponds to a location in the two dimensional sheet of paper with axes X and Y. Equivalently, we can specify the complex number by giving the distance r to the point and the angle θ made with respect to the X−axis. So the probability amplitudes which occur in quantum processes are specified by giving two numbers r and θ. Then the rule for taking the square of the probability amplitude and getting the actual chance for the process to take place is simply to compute r^2 and ignore θ. It is very important that the probability amplitude is a complex number; without this, we cannot get the correct kind of interference with electrons in the two-slit experiment. The key point is that probabilities (chances for processes to occur) are always positive; you cannot add a bunch of positive numbers and make them cancel each other — which is what we need in interference. On the other hand, the probability amplitudes can be positive or negative in the sense that x and y in Fig. 3.6 can take all possible values in the two dimensional plane, both positive and negative. This allows us to cancel one probability amplitude against another leading to interference.

How come the steel balls behave a lot saner than the electrons? The answer has to do with the wavelength that we associate with a particle. The rule is fairly simple. If the momentum of the particle is p, then the wavelength of the wave associated with that particle is about $\lambda = h/p$ where $h = 6.6 \times 10^{-27}$ erg sec is called Planck's constant, after the German scientist Max Planck who was one of the pioneers of quantum theory. If you take a steel ball of mass 1 gm moving with, say, 10 cm per second, the wavelength you will associate with it is about 10^{-28} cm which is extraordi-

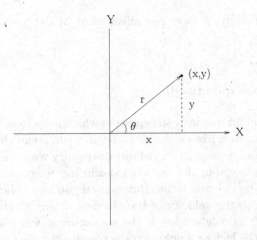

Fig. 3.6   The notion of a complex number. One can think of a complex number as a point in a two dimensional space given by a pair of real numbers (x, y). Equivalently, one can specify a complex number by giving the distance to the point from the origin (r) and the angle θ. More formally, a complex number z can be expressed in the form $z = x + iy$ where x and y are ordinary numbers and the symbol i stands for the imaginary quantity $i = \sqrt{-1}$.

narily small compared to even nuclear dimensions. On the other hand, an electron with a mass of about 10^{-27} gm moving with a speed of 10 cm per second, will have an associated wavelength of about 1 cm — much bigger than the size of the electron. As we saw earlier, in the case of light and sound, the longer the wavelength, the easier for you to observe the wave properties. With steel balls, you are never going to catch them as behaving like a wave and they will travel in straight lines just like light rays travel in straight lines when the wavelength is much smaller than other relevant dimensions. The reason we don't see quantum effects in every day life is essentially due to the very small value of Planck's constant h.

While the above description is basically correct, we have glossed over one important conceptual point. Suppose we arrange matters in such a way that we can actually determine through which hole the electron went or — more generally — which path an electron took. This will require setting a complicated array of detectors etc which will have to interact with the electron. It turns out that, if you do that, the electron will behave like a steel ball! So the behaviour of the quantum particle depends on whether you are observing it or not! This feature has created all sorts of debates and controversies about quantum theory and we will discuss it further in

Sec. 3.7. Fortunately, it does not affect most of the 'practical' results of quantum theory.

3.3 Things are quantized!

There are several interesting consequences which arise from our associating a wave with a particle like electron. To begin with, recall that Fig. 3.5 tells you that you cannot associate a definite trajectory with the electron. This is what we found even in the two slit experiment; we couldn't say through which hole a given electron went. But then, if you know the position of the electron as well as the velocity of the electron at any given moment, then one can trivially calculate what is the trajectory it will take immediately afterwards. So the lack of a unique trajectory for the electron also tells you that you cannot measure both the position and the velocity of an electron with arbitrary accuracy at any given instant. This result goes under the name *uncertainty principle* and was first stated clearly by the German scientist Werner Heisenberg. More precisely, this principle tells you the following: If you devise an experiment to measure the position of a particle at any given time with an accuracy α, its momentum will be uncertain by (at least) an amount $(h/4\pi\alpha)$. In other words, the process of measuring the position will disturb the system so that you cannot escape the resulting uncertainty in the momentum without sacrificing accuracy in the position measurement. If you try to measure the position with infinite accuracy (which corresponds to $\alpha = 0$) the momentum will be totally uncertain. You will know where the electron was at a given moment but will have no clue with what velocity it was moving! As a result, you wouldn't know where it will be at the next instant. Remember that, to associate a trajectory with a particle, you need to know both — where it is and how it is moving — simultaneously. Uncertainty principle tells you that you cannot do it thereby preventing you from associating a trajectory with a particle.

The second, and probably more important, result — at least from a practical point of view — has to do with the energy particles can have in different physical contexts. Since this is a key result in quantum theory, let us spend sometime understanding it.

Let us begin with a simple example of a particle which is enclosed in a box and moving horizontally all the time bouncing back and forth between two walls (see the left panel of Fig. 3.7). Suppose the particle has a mass m and speed v. It first travels to the right, say, hits the wall, bounces

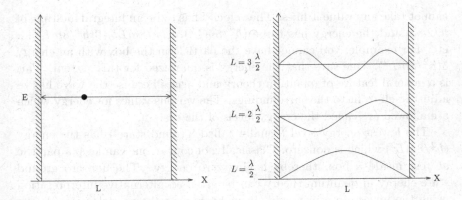

Fig. 3.7 Particle inside an one dimensional box in classical mechanics and quantum mechanics. (a) In classical physics, a particle is confined to a region along the x-axis, $0 < x < L$, can have any possible value for kinetic energy depending on its speed. It will bounce back and forth between the walls of the confining box located at $x = 0$ and $x = L$. (b) In quantum theory, the velocity of the particle is related to the wavelength associated with the particle. Unless the length of the box is equal to an integral multiple of half-wavelengths, the particle cannot be confined inside the box in a stationary manner. This leads to the quantization of the allowed energy levels of the particle.

elastically with the same speed, travels left to the other wall, bounces back etc. All the time it has the same speed v so that its kinetic energy is $E = (1/2)mv^2$. In principle, the particle can have any speed v and hence it can have any energy E.

The situation changes drastically when we associate a wave with the particle. Now it is necessary that the amplitude of the wave vanishes at $x = 0$ and $x = L$. This is because we know that the particle cannot be present outside the box at $x < 0$ or $x > L$; therefore the wave amplitude must vanish for $x < 0$ or $x > L$. Since the amplitude cannot jump in its value (the mathematical jargon being "it has to be continuous") in going from $x < 0$ to $x = 0$, it follows that the amplitude must vanish at $x = 0$. Similar reasoning tells you that it must vanish at $x = L$ as well. So we demand that the wave "fits" inside the box as shown in right panel of Fig. 3.7. This is possible only for certain wavelengths and not for all. The figure shows that, for the wave to fit in the box, the size of the box L should be an integral multiple of half the wavelength λ; viz., $L = n(\lambda/2)$ where $n = 1, 2, \ldots$ is any positive integer. Since the wavelength is given by $\lambda = h/mv$, it follows that $L = n(h/2mv)$ giving $v = nh/2mL$. The corresponding energy is $E = (1/2)mv^2 = n^2(h^2/8mL^2)$. We now see that the velocity or the energy

cannot take any value it likes! The velocity has to be an integral multiple of $h/2mL$ and the energy has to be $(h^2/8mL^2), 4(h^2/8mL^2), 9(h^2/8mL^2)$....
etc. For example, you cannot have the particle in the box with an energy $5(h^2/8mL^2)$. One says that the energy is quantized for this system. This is a general feature of quantum theory and arises because the wave has to adjust itself to fit to the surroundings. The various values for energy which are allowed are called the energy levels of the system.

The lowest energy level (usually called 'ground state') has the energy $(h^2/8mL^2)$ which is nonzero. Classically, of course, one can keep a particle at rest inside a box, thereby having zero energy. The nonzero ground state energy in quantum theory can be given an alternative interpretation using the uncertainty principle. If we know that the particle is inside the box (and we *do* know that), its position is known within accuracy of L. The uncertainty principle then tells you that the momentum has to be uncertain at least by the amount $h/4\pi L$. This momentum will correspond to an energy $p^2/2m = (h^2/32\pi^2 mL^2)$ which is roughly of the order of the ground state energy except for a numerical factor. A particle confined in a box of size L should have at least this much energy in quantum theory, because of the uncertainty principle. In this approach we could think of the excited, higher energy states as equivalent to confining the particle to a size smaller than L. If the wave has n nodes, we may think of the particle to be confined to a size L/n and the corresponding momentum (due to uncertainty principle) will rise proportional to n and the energy will scale as n^2.

Once again it is easy to understand why we don't see such effects in everyday life. If you have a ball of mass 10 gm bouncing between the walls of a box of size 50 cm, the energies are quantized in units of 2.2×10^{-58} ergs. That is, the energy difference between the levels with $n = 1$ and $n = 2$ is a tiny number 6.6×10^{-58} ergs. One will not see this in every day life and it will appear as though the energies are continuous.

Another important example with quantized energy levels which appears in different physical contexts is what physicists call a *harmonic oscillator*. The simplest example of a harmonic oscillator is a mass attached at the end of a spring. If you displace the mass slightly and let it go, we know that it will oscillate back and forth. A wide class of oscillations can be modelled by such a system in which the oscillating particle moves between two extremum positions with constant energy. The maximum displacement of the mass (in the above example) is called the amplitude of the oscillation. The energy of the system, when the mass is displaced by an amount x from

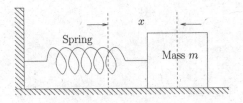

Fig. 3.8   An example of a simple harmonic oscillator in Newtonian mechanics is provided by a mass attached at the end of a spring. If the mass is displaced from the equilibrium position and let go, it will oscillate with a frequency ω determined by the properties of the spring.

its equilibrium position, can be expressed as the sum $E = K + V$ where $K = (1/2)mv^2$ is the kinetic energy of the mass m moving with speed v and $V = (1/2)m\omega^2 x^2$ is the potential energy provided by the spring (see Fig. 3.9). Here ω is the frequency of oscillation which is determined by the properties of the spring. When the mass is at the extreme point, its speed is zero and all the energy is given by the potential energy so that $E = (1/2)m\omega^2 a^2$. When the mass passes through the equilibrium position ($x = 0$) it will have maximum velocity and all its energy will be kinetic leading to the relation $E = (1/2)m\omega^2 a^2 = (1/2)mv_0^2$. We see that the maximum speed acquired by the mass is given by $v_0 = a\omega$. For a given spring and the mass, with given m and ω we can choose the amplitude of the oscillation a such that its energy can have any given value.

This classical situation gets changed by quantum theory. Once again, it turns out that the allowed energy levels for this system are quantized as shown in Fig. 3.9 and are given by $E_n = \hbar\omega(n + 1/2)$ where $n = 0, 1, 2, \ldots$ and $\hbar$ (called 'hbar') is a convenient notation for $h/2\pi$. There are two important features which are worthy of note in this example. First the separation between energy levels is a constant and is equal to $\hbar\omega$; that is, the energy levels are equidistant. Second, the lowest energy level, viz. the ground state, again has a nonzero energy given by $E_0 = (1/2)\hbar\omega$. Classically you could have attached a mass to a spring and could have left it alone without elongating or compressing the spring. The mass will not move, will have zero energy and everything will be fine. This would correspond to a particle being located at $x = 0$ in the potential well in Fig. 3.9. But you will recall that specifying the position of the particle as well as its speed simultaneously is forbidden by uncertainty principle. So you cannot say, in quantum theory, "the particle is located at $x = 0$ and is not moving". When one tries to confine the particle by a potential,

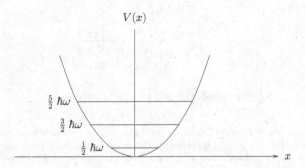

Fig. 3.9   The harmonic oscillator in quantum theory. While an oscillator can have any energy in classical theory, it can have only discrete energy levels in quantum theory. The lowest energy level has the energy $(1/2)\hbar\omega$ in quantum theory while classically the lowest energy state has zero energy. The excited states differ by equidistant steps of $\hbar\omega$.

the uncertainty principle will give it certain amount of momentum and an associated kinetic energy. If you confine the particle too much, the price you pay through uncertainty principle will be substantial. So the ground state of the system is obtained as a compromise between the tendency of the potential to confine the particle and the tendency of the uncertainty principle to increase its momentum. It is this compromise which again leads to a nonzero energy for the ground state of the system.

Vibrating systems described by a harmonic oscillator is ubiquitous in nature. In Chapter 1, Sec. 1.6 we pointed out that two of the major problems faced by classical physics was in describing the specific heats of gases and the radiation energy density in a blackbody cavity. The way quantum theory solves these difficulties is intimately related to the quantization of energy levels of a harmonic oscillator. Consider, for example, the vibrational mode of a diatomic molecule. Classically, the energy stored in vibration can take any value and one can excite the vibration by giving arbitrarily small amount of energy to the molecule. Quantum mechanically, the vibrating system will be described by a harmonic oscillator and its energies will be quantized. To excite the system from one energy level to another you need to supply a *minimum* energy which corresponds to the spacing of energy levels — viz., an amount $\hbar\omega$. In certain contexts, like for example at low temperatures, it may not be possible to supply this amount of energy to the system. Then quantum mechanically the vibrational modes cannot be excited and will not contribute to specific heat. This is the primary reason why the classical calculation of specific heat based on counting all the degrees of freedom fails at low temperatures. The situation regarding radiation density in a blackbody cavity is similar. The exchange of energy between the radiation and matter forming the walls of the cavity has to be in quantized units which makes the description different when the temperature is not high enough to contribute sufficient energy to excite the modes. This again leads to a wrong description in classical theory (which assumes that energy is not quantized) compared to quantum theory.

Another example of the same principle in operation is in the case of the hydrogen atom. Recall that, according to classical electromagnetic theory, an electron orbiting a proton in the hydrogen atom is undergoing an accelerated motion and should emit electromagnetic radiation. If that happens the electron will quickly loose energy and spiral down to the nucleus and the atoms will be unstable. Quantum theory ensures stability by modifying this description in two essential ways. First, it prevents us from talking about the trajectory for the electron. Second, uncertainty principle prevents us from placing an electron too close to a nucleus. If the electron is localized within a radius r of the proton in a hydrogen atom, uncertainty principle requires it to have a momentum of at least the amount $p = \hbar/r$ and a corresponding kinetic energy $K = p^2/2m = \hbar^2/2mr^2$. On the other hand, when it is at a distance r from the proton, it has an electrostatic potential energy $V = -e^2/r$ where e is the electric charge of the electron. The total energy is $E = K + V = \hbar^2/2mr^2 - e^2/r$. The electron tries to minimize this energy. In the absence of quantum theory, uncertainty principle etc., the first term is absent and the energy can be made as low as you want by decreasing r. Because of the first term, we see that this strategy will not work; if you decrease r too much, the first term will start dominating and will make the energy go up. So there is a particular radius r_0 at which the above expression takes the minimum possible value and this is what determines the stable ground state of the hydrogen atom. The existence of a ground state is what gives stability to atoms and thus to all matter.

In this case as well, there are excited states with quantized energy levels. To obtain them, it is best to consider another important quantity which is quantized for any physical system, viz., the angular momentum. Because of its importance, we shall describe it in some detail. Angular momentum of a rotating body, in Newtonian mechanics is essentially the sum of the momenta of the particles in the body but weighted by the distance of the particles from the axis of rotation. (This was described in Chapter 1; see Fig. 1.5 and Fig. 1.6.) If a particle of mass M moves in a circular path of radius R with velocity v, then its angular momentum is MvR. The angular momentum is a vector and — in this case — it is directed along the direction of the axis of rotation. You see that one can increase the angular momentum by increasing any of the three quantities M, v or R. Its importance stems from the fact that, for an isolated system, the angular momentum is a conserved quantity, just like energy or momentum.

Classically, the MvR of an orbiting particle can have any value. But not so quantum mechanically. The orbital angular momentum of a system in quantum mechanics can only take values which are quantized in units of $\hbar \equiv h/2\pi$, which — as you can verify — has the dimensions of angular momentum. This result is applicable, for example, to an electron orbiting a proton in the hydrogen atom. In this case, one can give this result a rather curious interpretation. We saw earlier that, one can associate with a particle of momentum $p = mv$ a wavelength (h/p). If such particle is moving in a circular orbit of radius R, it seems reasonable to assume that one should be able to fit an integral number of these wavelengths along the circumference (see Fig. 3.10). This will require $2\pi R = n(h/mv)$. It follows that $mvR = n(h/2\pi) = n\hbar$ thereby leading to the quantization of the orbital angular momentum. [Though we have drawn a nice picture in Fig. 3.10, it is not to be taken literally since trajectories are not well defined in quantum theory!]

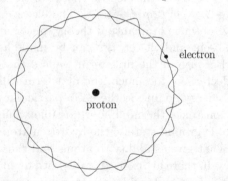

Fig. 3.10   Quantization of the orbital angular momentum for an electron orbiting a proton. The wavelength associated with the electron is assumed to have an appropriate value so that it fits evenly into the circumference. This requires the orbital angular momentum to be an integral multiple of $\hbar$.

From this result, with some algebra, it is possible to show that the energy of the electron in the hydrogen atom is also quantized. The lowest energy it can have, when it is most tightly bound to the proton, which is the ground state energy, corresponds to $E_R = -13.6$ eV. (It is a common practice to add a minus sign to the energy in order to show that the electron is bound inside the atom. By convention, a free electron at rest has zero energy and a free, moving electron has positive energy. We know that we have to supply 13.6 eV of energy to "free" the electron in a hydrogen atom from the grasp of the proton. So the electron in hydrogen atom must have *less* energy than the free electron — which, by convention, has zero energy; so, the energy of the electron bound in an atom should be indicated by a negative number). In addition to the ground state, the electron can also exist at several higher energy levels ("excited states") with energy $E_n = -|E_R|/n^2$. The first excited state has an energy of $-[13.6/(2 \times 2)]$ eV $= -3.4$ eV and the second excited state has an energy $-[13.6/(3 \times 3)]$ eV $= -1.5$ eV etc.. Note that -3.4 is *less* negative than -13.6; so -3.4 is larger than -13.6. Thus an electron with energy -3.4 eV has higher energy than an electron with energy -13.6 eV.

There is one more aspect about energy in quantum theory that deserves mention. We saw earlier that the uncertainty principle prevents us from measuring both the position and speed of a particle simultaneously. If we try to measure the position of the particle with an accuracy ϵ, say, it will inevitably lead to an uncertainty in the speed of the particle given by $\hbar/m\epsilon$ where m is the mass of the particle. It turns out that there is a corresponding uncertainty relation connecting the energy with time. If you try to measure the energy of a system within a time interval τ, then the result will be uncertain at least by an amount $\hbar/\tau$. This plays a key role in certain aspects of quantum field theory which we will discuss later in Chapter 4.

3.4 Spin and the social behaviour of particles

As usual this is not the whole story about the angular momentum and the mystery deepens when you probe the situation further. To begin with recall that the angular momentum is a vector. If we take the component of the angular momentum of the system along some axis (say z axis) it has to be quantized in units of $\hbar$; but the weird thing is that if we change our mind about which direction is z-axis and take a new z-axis which is inclined at, say, 7 degrees to the original axis, the angular momentum will be quantized along this axis too! Some thought will convince you that this cannot happens for usual 'classical' vectors. The reason it works in quantum mechanics is again related to the issue of measurement. It turns out that you cannot simultaneously measure the angular momentum along the z-axis and, say, the x-axis. It is this feature which allows one to escape contradictions in quantum theory.

The second mystery with angular momentum is a still deeper. You might have noticed that, talking about the electron moving around the proton in a hydrogen atom, we specifically used the term *orbital* angular momentum. This is because the particles (like electrons) also have an *intrinsic* angular momentum called 'spin'. The spin of a particle is also quantized, except that the spin is quantized in units of $\hbar/2$ [not $\hbar$]. The electron always has the spin $\hbar/2$ and one usually says that electron is a spin-1/2 particle. Each particle carries a specific value for its spin; the proton also has spin-1/2 while the photon (treated as the quanta of light) has spin-1.

One might be tempted to think of spin of the electron as something due to its intrinsic rotation: as though electron is a tiny ball which is rotating on its own axis and thereby having an intrinsic angular momentum. Unfortunately, this interpretation runs into all kinds of problems and hence has to be abandoned. One of the reasons is that — just like orbital angular momentum — the spin of the electron is $\pm(\hbar/2)$ along *any* axis; this is hard to explain if we imagine electron to be a tiny rotating ball. In fact, nobody has a deeper understanding of electron spin than what is described above! (Mathematically sophisticated description, yes; deeper understanding, no.) With the addition of spin, we see that a particle like the electron has 3 different attributes: its mass, its charge and its spin. This is true more generally in quantum theory. Any given particle can be described by giving a set of numbers which uniquely characterizes its behaviour.

The spin of a particle plays a crucial role in its behaviour. Physicists classify particles as either *bosons* or *fermions* based on the spin. Particles with half integral spin (1/2, 3/2, 5/2 ...) are called fermions and particles with integral spins (0, 1, 2, ...) are called bosons. The key feature which distinguishes fermions from bosons is the following. *No two fermions can exist in the same quantum state simultaneously.* On the other hand, bosons show an enhanced preference to be in the same state. This innocuous looking result is very deep and its rigorous proof requires advanced concepts from relativistic quantum theory. In fact, but for this result — that no two fermions can exist in the same state — there will be no periodic table, no chemistry and possibly no life!

To see this, consider how the structure of an atom changes as we increase the number of protons and the number of electrons in it. The simplest atom, that of hydrogen, has one proton and one electron orbiting around it and the electron can be put in the lowest energy state of the atom. Consider next the helium atom with two electrons. Since electrons have spin half, they are fermions and no two of them can occupy the same quantum state. So it will be impossible to put both the electrons in the ground state of the helium atom except for the fact that the spin of each electron can be either up or down along a given direction. This allows us to put the first electron in the ground state with its spin up and the second electron in the ground state with its spin down. Because the spins are different, these states are actually different with respect to the angular momentum and hence you are allowed to keep both the electrons in the ground state.

Consider next the lithium which has 3 electrons. You put the first two electrons in the ground state with one spin up and one spin down. The third one cannot now be put in the ground state with any value for the spin! This means you mandatorily have to keep the third electron at the first excited state of the lithium atom. The same kind of rules decide the lowest energy configuration for any other atom with larger number of electrons. It is this fact — that not all the electrons can be lumped into the lowest energy ground state — which makes the chemical properties of various elements to be different and leads to the so called periodic table of elements.

But why is it that there are fermions and bosons with such different behaviour? The rigorous reason, of course, involves fairly sophisticated mathematical arguments but it is possible to provide a qualitative explanation along the following lines. Ultimately, this result is linked to the concept of *indistinguishability* of particles in quantum theory which we will first describe. In classical physics, a particle moves from event to event along a definite trajectory. This means that, if two particles move from two initial points to two final points, we will be able to determine which particle went where. But in quantum theory, particles do not have well defined trajectories and we can only say that there is a chance for two particles to end up at two points X and Y at some time $t = t_2$ if there were two particles at point A and point B at an earlier time $t = t_1$. You can no longer say whether the particle which was at point A ended up at point X or at point Y. This brings about an indistinguishability with respect to the exchange of particles in the quantum description and the quantum states which are allowed for the particles must respect this. Let us try to describe this in a somewhat more quantitative manner. For this, we need to understand a couple of more key principles of quantum theory.

First, let us consider two particles with the first particle is in a state described by the wave amplitude $\psi(1)$ and the second particle is in a different quantum state described by the amplitude $\phi(2)$. How do we describe a state having both particles? If the particles are distinguishable — say, one is an electron and the other is proton — and do not influence each other strongly, then we can take the state describing both the particles together as given by the product $\psi(1)\phi(2)$. In other words, we can multiply the wave amplitudes to describe the simultaneous occurrence of two independent possibilities. This will make sense to you when you recall that wave amplitudes are somewhat like probabilities. If you roll a die, the chance C_1 for getting the number 6 is just $C_1 = (1/6)$. If you toss a coin, the chance C_2 of getting heads is $C_2 = (1/2)$. If you now roll a die *and* toss a coin, the chance for getting a 6 as well as heads is, of course, the product $C_1 C_2 = (1/12)$. It is exactly this product rule we are using for wave amplitudes.

But the situation changes when the particles are indistinguishable. In this case, one cannot describe the state of both the particles by the product $\psi(1)\phi(2)$. This is because, if we interchange the particles we end up getting the state $\psi(2)\phi(1)$ which will, in general, have completely different physical properties. This will make the two particles distinguishable which we do not want. To solve this problem, we have to make use of another property of quantum theory known as *superposition principle*. This principle says that, if a quantum system can exist in a state represented by an amplitude Ψ or in a state represented by an amplitude Φ, then it can also be in a state represented by the states $\Psi + \Phi$ or $\Psi - \Phi$. (Actually, the superposition principle is more general but this version is adequate for our purpose.) In our case, since the quantum state can be represented either by $\psi(1)\phi(2)$ or by $\psi(2)\phi(1)$, we can also describe the state representing the two particles simultaneously as either the symmetric state $S(1,2) = \psi(1)\phi(2) + \psi(2)\phi(1)$ or the antisymmetric state $A(1,2) = \psi(1)\phi(2) - \psi(2)\phi(1)$. In the state $S(1,2)$ the interchange of two particles has no effect: $S(2,1) = S(1,2)$. On the other hand, in the antisymmetric state, the interchange of particles leads to an extra minus sign: $A(2,1) = -A(1,2)$. This is actually as good as nothing changing because the probabilities in quantum mechanics are computed by taking the square of the amplitude.[1] Since $A^2(2,1) = A^2(1,2)$ there is no harm done even when the interchange of particles leads to a

---

[1] We are simplifying matters somewhat here but keeping the essence intact. Actually the amplitudes are complex numbers and taking their squares is a little bit more involved.

minus sign. So the quantum mechanical description of a pair of identical particles should use either $S(1,2)$ or $A(1,2)$.

Consider now what happens when the particles are in the same state so that $\psi = \phi$. For the symmetric state we get $S(1,2) = 2\psi(1)\psi(2)$ which shows certain enhancement in the amplitude but nothing very dramatic happens. On the other hand, look at what happens to the antisymmetric state when $\phi = \psi$: $A(1,2) = \psi(1)\psi(2) - \psi(1)\psi(2) = 0$! In other words, the amplitude for the two particles to be found in the same state identically vanishes — which is the same as saying you cannot put two particles in the same quantum state if the quantum state has to be antisymmetric. Bosons are those particles for which the combined state is described by a symmetric amplitude while fermions are particles for which the combined state is described by an antisymmetric amplitude.

The above description gives a rough idea as to why indistinguishability of particles in quantum theory leads to the existence of bosons and fermions. What is more involved is to justify why particles with half integral spin must be described by the antisymmetric amplitude while the particles with integral spin must be described by the symmetric amplitude. This feature, as we said before, requires fairly sophisticated tools to prove rigorously.

3.5 Interaction of matter and radiation

A good example of quantum theory in action, in a very practical sense of the term, is provided by the interaction of radiation and matter. This is also an area which led to several early insights in the development of quantum theory and hence deserves some discussion.

To understand how radiation is emitted and absorbed by matter, we have to again start with the intrinsic wave-particle duality in the description of nature. Originally we talked of electromagnetic radiation as a wave and electrons, for example, as particles. But as we just saw, we can associate a wave nature with the electron through the rule $\lambda = h/p$ which relates the wavelength to the momentum. In a complementary sense, it turns out that, one can think of electromagnetic radiation as made of particles called photons. The momentum of the photon p is now related to the wavelength of the radiation by $p = h/\lambda$. But we saw in the last chapter that the energy of a relativistic particle is related to its momentum in a specific way. For a particle like photon, moving with speed of light, this relation is given by $E = pc$. Hence the energy is related to the wavelength by

$E = pc = hc/\lambda$. Using the fact that for the radiation, $\lambda\nu = c$ we get the simple relation between energy of the photon and the frequency of the electromagnetic radiation: $E = h\nu$. The energy of an individual photon depends on the frequency of the radiation and the higher the frequency, the higher the photon energy. An individual photon of wavelength of 5000 Å has a frequency of 6×10^{14} Hz and an energy of 4×10^{-12} ergs which is about 2.5 eV. One can think of electromagnetic radiation either as a wave or as a bunch of photons. A beam of light can carry more energy either by having more photons of the same energy or by having the same number of photons each of which has more energy.

We also know that the wave nature will make its presence felt only when the wavelength (which is related to the energy) becomes comparable to other length scales like, for example, the size of the measuring apparatus, while the particle nature will be apparent when the wavelength is small compared to other length scales. Because of this reason, long wavelength electromagnetic radiation — like radio waves, infrared or even visible light — can be usually thought of as a wave; at the other extreme, short wavelength radiation — like X-rays or gamma rays — is usually considered as made up of photons.

At the tiny scales of atomic systems, we have to take into account the particle nature of light. In fact, atomic systems emit and absorb radiation in the form of photons. Consider, for example, the electron orbiting the proton in a hydrogen atom. As we saw above, such an electron cannot have any arbitrary amount of energy while in the hydrogen atom and the energy levels are given by $E_n = -13.6$ eV$/n^2$ where $n = 1, 2, 3....$ etc. It is, however, possible for an electron to jump from a higher energy level to a lower energy level on its own, changing its energy. When an electron jumps from the first excited state ($n = 2$; which has an energy of -3.4 eV) to the ground state ($n = 1$; with an energy of -13.6 eV), its energy changes by the amount 10.2 eV which is the difference in these energies. This energy is emitted as a photon with a frequency of 2.5×10^{15} Hz (recall that $E = h\nu$) and a wavelength of about 1216 Å. This is an example of the mechanism by which a hydrogen atom can emit electromagnetic radiation. Similarly, by making transitions between different energy levels, the atom can emit radiation of different wavelengths (see Fig. 3.11).

The same physical process is responsible for the emission of radiation in almost all cases. Different atoms will have different sets of energy levels and the transitions of electrons between these levels will lead to the emission of radiation at different frequencies. In fact, the process is not restricted to

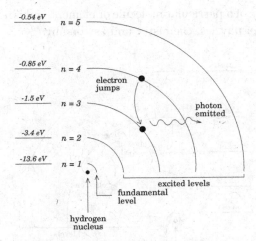

Fig. 3.11   An electron orbiting the nucleus can exist only with certain definite values for the energy. The lowest energy level for an electron for a hydrogen atom has an energy of −13.6 eV. (This is the level marked n = 1 in the figure and corresponds to the electron being closest to the nucleus.) The next energy level marked n = 2 has an energy of −3.4 eV. This is a higher energy level because it is less negative. The figure shows the first 5 energy levels in the hydrogen atom. When an electron 'jumps' from a higher level to a lower level, it emits the difference in the energy in the form of a photon. This photon will have a wavelength characteristic of the energy difference between the two levels.

atoms. Molecules, for example, have their own energy levels (in addition to the energy levels of the atoms which make up the molecule) and hence can emit radiation by making a transition between two such levels. More importantly, an electron in a given energy state does not emit any radiation which is in stark contrast with the classical situation in which an electron in a circular orbit has to emit radiation because it is in a accelerated trajectory. As we said before, quantum theory escapes contradictions because one cannot attribute a trajectory to the electron.

The pattern of radiation emitted by a particular atom (or molecule) is called its "spectrum" or, more precisely, "emission spectrum". Since the actual spacing between the energy levels differ from atom to atom (and molecule to molecule) the spectrum of hydrogen will be quite different from, say, the spectrum of carbon dioxide. By studying the spectrum of some source of radiation, one can often pin-point the elements which are present in that source. What is more, by measuring the amount of energy that is emitted in different transitions, one can also determine the relative abundance of elements which are present. In this way the spectrum serves like the signature of a particular molecule or element and is a very important diagnostic tool in physics, chemistry and astronomy.

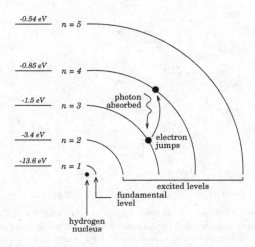

Fig. 3.12   This figure shows the absorption of a photon, just as Fig. 3.11 showed the emission. The electron can absorb a photon and jump from a lower level to a higher level provided the photon has exactly the energy corresponding to the energy difference between the two levels. If the energy of the photon is different, then it will not be absorbed by the hydrogen atom.

The absorption of radiation proceeds in a similar manner but in the opposite direction. Consider, for example, an electron in the ground state of hydrogen atom. It cannot jump to a higher energy state on its own since this process would require some energy. But if radiation is shining on the atom, it can absorb a photon of wavelength 1216 Å and make a transition from the ground state to the first excited state; or it can absorb a photon of wavelength 1025 Å and make a transition to the second excited state and so on (see Fig. 3.12). Other atoms and molecules with different sets of energy levels behave in a similar manner; all of them can absorb photons and make transitions from lower energy levels to higher energy levels.

Just as in the case of emission, an atom (or molecule) can only absorb photons of definite wavelengths when it makes a transition between two energy levels. Suppose we shine light of all wavelengths on a system made of a large number of hydrogen atoms. Photons of most wavelengths will pass through the gas while photons of specific wavelengths (like 1216 Å and 1025 Å) will be absorbed by the atoms (see Fig. 3.12). On observing the radiation which has passed through the hydrogen gas, one will clearly see the tell-tale absence of photons at particular wavelengths. Such a spectrum is called absorption spectrum. By studying an absorption spectrum we can determine the composition of elements in regions through which radiation has been passing through.

The above description of absorption and emission of radiation is based on the picture of photons. Since radiation can be thought of either as photons or as waves, you may wonder how the same phenomena can be described in terms of waves. For example, are radio waves made of photons? At a basic level the answer is, of course, 'yes'. But as we said before, if the wavelength of radiation is comparable to other length scales involved in the problem, then it is better to think of electromagnetic radiation as a wave. For visible light, the wavelength is like a 50 millionth part of a centimeter; clearly this is tiny compared to everyday scales. Radio waves, in contrast, could be tens or hundreds of meters in wavelength which could be comparable or bigger than other length scales in the system. In that case one may think of electromagnetic radiation as "waves" which are being emitted by the motion of electrically charged particles. Whenever a charged particle moves with a variable speed, its kinetic energy changes and the difference in kinetic energy could be radiated in the form of electromagnetic radiation. You would have noticed that in any process involving the emission of electromagnetic radiation the energy of the charged particle changes. In the photon picture, this arises because the system emitting the radiation makes a transition from one energy level to another. In the wave picture this arises because the kinetic energy of the charged particle changes.

3.6 The atomic nuclei

At the atomic level, all matter is made of protons, neutrons and electrons. One can classify all matter by giving the number of protons (which is called the atomic number) and the total number of neutrons and protons in the nucleus (called the atomic weight, since this number tells us how heavy the atom will be).

Each chemical element has a unique atomic number starting from 1 for hydrogen, 2 for helium etc. The chemical properties of the element are essentially decided by the number of electrons in the element which, of course, is equal to the number of protons in the nucleus. The atomic weight of an element, however, is not quite unique. Take, for example, hydrogen has only one proton in the nucleus and hence has an atomic weight of unity. There exists another element called deuterium which has one proton and one neutron in its nucleus (and one electron orbiting around it). Clearly, both deuterium and hydrogen have the same number of protons (and electrons) and hence the same atomic number; but the atomic weight of deuterium is 2 (because it has a proton *and* a neutron in the nucleus) while that of hydrogen is 1. Since the chemical properties of the elements are determined by the number of electrons, both deuterium and hydrogen have very similar chemical properties. (For example, two atoms of deuterium can combine with one atom of oxygen and form "heavy water", denoted by D_2O, just as two atoms of hydrogen can combine with one atom of oxygen and form ordinary water, H_2O.) Similarly, normal helium has an atomic number of 2 (two protons and two electrons) and atomic weight of 4 (two protons and two neutrons in the nucleus). However, there exists another species called helium-3 which has an atomic number of 2 (two protons and two electrons) but an atomic weight of 3 (two protons and *one* neutron in the nucleus). Once again the chemical properties of normal helium (which may be called helium-4) and helium-3 are quite similar. Such species — like hydrogen and deuterium, or helium-3 and helium-4, — are called "isotopes". Though the chemical properties of isotopes of the same element will be similar, their atomic nuclei are quite different and behave differently.

As the atomic number increases, the atoms become bigger and, in general, exhibit complex properties. Naturally occurring elements have atomic numbers ranging from 1 to 92. For example, aluminum has an atomic number of 13 and atomic weight of 27. This means that an atom of aluminum contains 13 protons (and, of course, 13 electrons). Further, since the atomic weight is 27 and atomic number is 13 there must be $27 - 13 = 14$ neutrons in the nucleus (see Fig. 3.13).

All the elements do not occur with equal abundance in the universe. It turns out that nearly 90 percent of the matter in the universe is made of hydrogen and 9 percent is helium. The remaining 100 odd elements, called heavy elements, make up only about 1 percent of the matter in the universe. (Of course a planet like the Earth has a lot more of heavy elements; but the Earth is not a typical example for computing cosmic abundance.)

The structural properties of an atom, like its binding energy, are determined by the electromagnetic interaction between protons in the nucleus and electrons which orbit around it. Thus all physical and chemical properties of gross matter in its four states (solid, liquid, gas and plasma) are

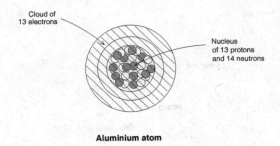

Aluminium atom

Fig. 3.13 Normal atoms are made of a nucleus consisting of protons and neutrons with electrons orbiting around the nucleus. The protons carry positive electric charge and electrons carry negative electric charge. Normally, atoms have an equal number of protons and electrons and hence are electrically neutral. The properties of the various elements are determined by the number of protons and neutrons in the nucleus. The figure shows (schematically) an aluminum atom which has 13 protons and 14 neutrons in its nucleus with a cloud of 13 electrons orbiting around it. Note that the figure is not drawn to scale; the size of the nucleus is highly exaggerated.

governed by the electromagnetic force. The electromagnetic force, however, does not play the major role in the structure of the atomic nucleus itself. As we saw before, the nucleus of aluminum contains 13 protons and 14 neutrons. Since protons are all positively charged particles and like charges repel each other, the nucleus will be blown apart if there is no other force to hold the protons together. Indeed there exist forces other than electromagnetic force, which are responsible for the stability of atomic nucleus.

These forces, called "strong force" and "weak force", have very different characteristics. The strong force is considerably stronger than the electromagnetic force and provides the energy for binding the nucleus together. A typical atomic nucleus has a binding energy which varies from 7 million electron volt to 500 million electron volt. To describe such nuclear processes, we will use the unit "MeV" which stands for million electron volts. So it is considerably stronger than the electromagnetic force. That is, two protons which are sufficiently close together will feel a nuclear force which is much stronger than their electrostatic repulsion. But the strong force is felt only when two particles are very close to each other. If two protons are separated by a distance of about 10^{-13} cm then the strong force is 100 times larger than the electrical force. But if the protons are moved apart to 10^{-12} cm then the strong force decreases by a factor of about 10^5 while the electrical force decreases only by 10^2. So, as the distance increases,

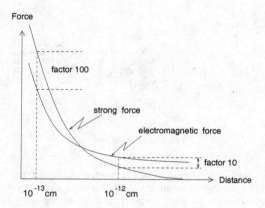

Fig. 3.14 This schematic diagram shows the variation with distance of the strength of electromagnetic and strong force between two protons. The strong force is attractive while the electromagnetic force is repulsive. At very short distances the strong force dominates over the electromagnetic force and the protons can be bound to each other. The strong force, however, decreases rapidly with distance while the electromagnetic force falls only slowly with distance. As a result, at distances larger than about 10^{-12} cm the electromagnetic force will dominate over the strong force.

strong nuclear forces tend to be ineffective compared to electrical forces (see Fig. 3.14).

We saw that electrical forces exist only between particles carrying electric charge and neutral particles are unaffected by them. On the other hand, protons and neutrons feel strong nuclear force while electrons do not. One may say that protons and neutrons carry "strong charge" (just as protons carry "electric charge") while electrons are "neutral" as far as strong nuclear forces are concerned. In general, physicists classify elementary particles as "leptons" and "hadrons" of which leptons do not feel the strong force while hadrons do. Neutrons and protons are examples of hadrons while the electron is a lepton.

In addition to the strong nuclear force, the elementary particles also feel a weak nuclear force. This force has similarities to both strong and electromagnetic forces but, at the same time, has some vital differences. To begin with, it is about 10^{11} times weaker than the electromagnetic force. It also has a much shorter range than the electromagnetic force. In fact, its range is comparable to that of the strong force. But, unlike the strong force, both leptons and hadrons are affected by the weak force. In reality the weak nuclear force and the electromagnetic force are only two facets of a single entity called electro-weak force. When the interacting particles have very high energies (more than about 50,000 MeV), they feel the unified effect of the electro-weak force. But at lower energies, this force manifests itself as two separate forces. (We shall say more about this in Chapter 4.)

The binding energy varies from nucleus to nucleus in a fairly complicated manner. On the whole, the binding energy (per nucleon) increases with atomic number up to iron and then decreases, making iron the most stable of all nuclei. This suggests that elements heavier than iron will behave differently, compared to those lighter than iron, as far as nuclear processes are concerned. In addition to this average behaviour, one can also see significant spikes at some particular elements. For example, helium has an extremely stable nuclear structure, with a fairly high binding energy.

Since the nuclear forces provide the binding for the constituents of the atomic nucleus, we can split the atomic nucleus by overcoming these forces. The situation here is quite similar to the case of hydrogen molecule or hydrogen atom. By colliding two molecules or two atoms with sufficiently high speeds, we could supply enough energy to overcome the molecular or atomic binding. Similarly, by colliding two nuclei together at very high speeds we can provide the necessary energy to overcome the nuclear binding. In such — sufficiently energetic — collisions we can actually split the nucleus of an atom. For example, if we hit the nitrogen nucleus with the nucleus of helium, we can obtain two different nuclei; namely, hydrogen and an isotope of oxygen (see Fig. 3.15). Originally, the nitrogen nucleus contained 7 protons and 7 neutrons while the helium nucleus had 2 protons and 2 neutrons. In the final state, the isotope of oxygen has 9 neutrons and 8 protons while the hydrogen nucleus is made of 1 proton. Notice that we have now broken new ground. By splitting the nucleus in such a high energy collision, we have actually converted nitrogen and helium into oxygen and hydrogen. If sufficiently high energies are available, it is indeed possible to convert one element into another. By processes like this, one can actually split a nucleus into two; one calls this process "nuclear fission".

It is also possible to fuse two nuclei together and form a new element by a process called "nuclear fusion". For example, consider the element deuterium, which has an atomic nucleus made of one proton and one neutron. By colliding two deuterium nuclei one can produce a nucleus of helium. But since all atomic nuclei are positively charged and like charges repel, the nuclei of two deuterium atoms will have a tendency to repel each other. To fuse these two nuclei together, one has to overcome the repulsion by colliding the nuclei at very high energies. At such high energies, the nuclei can come very close to each other and — at such close distances — the

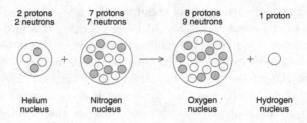

Fig. 3.15   The collisions between nuclei can lead to the transformation of one element to another. For example, this figure shows the nuclear reaction in which a helium nucleus (with 2 protons and 2 neutrons) collides with a nitrogen nucleus (with 7 protons and 7 neutrons) and forms an isotope of oxygen nucleus (with 8 protons and 9 neutrons) and a hydrogen nucleus made of a single proton. Such nuclear reactions can absorb or release large quantities of energy.

nuclear attraction can overcome the electrostatic repulsion. It follows that nuclear fusion can take place only when the particles are moving with very high speeds, that is, at very high temperatures.

Fission and fusion of nuclei involve changes in energy. As we saw before, the binding energy is different for different elements. Suppose we fuse the nuclei of A and B and obtain the nuclei of C. If the binding energy of C is lower than the combined binding energy of A and B then this process will release the difference in the energy. On the other hand, if C has higher binding energy, then one needs to supply energy to trigger this process. Whether a given process will release or absorb energy depends on the atomic number with the transition occurring at the atomic number of 56, which corresponds to iron. We saw earlier that iron has the maximum binding energy per nucleon so that the binding energy curve is sloping downwards to the right of iron. In this region we can break up an element to form *lighter* elements which have higher binding energy, i.e., are more stable. Such a process is preferred by nature and elements heavier than iron — in general — release energy when split up. The situation is reversed for elements which are lighter than iron. Here we can increase the binding energy by fusing two light elements together. Once again such a process is preferred and energy will be released in the nuclear fusion of lighter elements. In summary, we may say that heavier elements like uranium will release energy when undergoing fission while lighter elements like hydrogen will release energy during fusion. Of course, in order to fuse two nuclei together, we have to overcome the electrostatic repulsion and hence such a process can take place only at very high temperatures.

The processes described above form the basis of release of nuclear energy both in nuclear reactors as well as in nuclear bombs. The hydrogen bomb, for example, is based on the fusion reaction. Since fusing two hydrogen nuclei can release about 0.4 MeV of energy, one gram of hydrogen (containing 6×10^{23} atoms) undergoing fusion will release an energy of about 10^{17} ergs. For comparison, one gram of coal, when burnt, produces only 10^{11} ergs of energy, showing how vastly more powerful nuclear processes are. At a fundamental level, this difference is due to the difference in strengths between nuclear and electromagnetic forces. The burning of coal is a chemical reaction governed by electromagnetic forces, while the fusion of hydrogen is governed by nuclear forces.

There is another curious physical process which takes place in some of the heavy atomic nuclei which is inherently quantum mechanical and worth discussing. This process, called radioactivity, involves the spontaneous decay of the nuclei emitting energetic particles. This is divided into three separate processes called alpha decay, beta decay and gamma decay.

We will first describe alpha decay, involving an atomic nuclei spontaneously emitting an alpha particle and changing itself. The alpha particle is just the nuclei of helium atom and is made of two protons and two neutrons. The question is: Why does some nuclei behave in this strange fashion throwing out two protons and two neutrons? To understand this, let us again go back to the question of what holds the nuclei together. When two protons are brought very close to each other, the attractive strong interaction force dominates over the electrostatic repulsion which allows the nucleons to be bound together. This idea can be expressed in an equivalent fashion by calculating the energy of interaction between the two protons as a function of the separation between them. This interaction potential energy, as it is called, is shown in Fig. 3.16. At large distances, the interaction energy is contributed essentially by electrostatic repulsion while at small distances it is contributed by the strong interaction force. It is conventional to think of interaction energies as negative when the force is attractive and as positive when the force is repulsive. The attraction due to the strong force dominates at short distances, $0 < x < R$ (so that the potential energy is negative) while the electrostatic repulsion dominates at larger distances (with potential energy being positive). Now consider a particle of some energy E located at different distances. If such a particle is moving from a large distance towards the origin (along the part AB of the figure) then its kinetic energy will keep decreasing as it moves towards the origin. This is easy to understand because the total energy E is a conserved quantity and

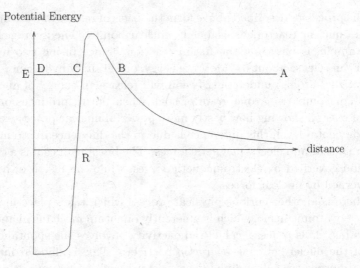

Fig. 3.16   The tunnelling of a particle through a potential barrier and the explanation of α decay of a nuclei. In Newtonian mechanics, a particle with energy E (marked by the horizontal line) which is located in the region CD cannot go to the region AB, To do so, it has to cross through the region CB in which the potential energy is larger than the energy of the particle. This is called a potential barrier which confines the particle. In quantum theory, it is possible for a particle to go from the region DC to region BA through a process called tunnelling.

is equal to the sum of kinetic and potential energies. As one moves from A to B, the potential energy increases along the curve and since E is a constant, its kinetic energy will decrease. At B the potential energy is equal to the total energy E and hence its kinetic energy will be zero. In other words, a particle with energy E cannot come any closer than the point B because by the time it reaches B its kinetic energy — and thus its speed — would have been reduced to zero. Similar considerations apply to a particle which finds itself in the region CD with energy E. It can move freely between C and D but at C its total energy is equal to the potential energy making the kinetic energy zero. In the region between C and B, the potential energy is higher than the energy which the particle has; therefore such a particle can never be found in the region CB. We say that the particle with energy E faces a potential barrier. If it was originally in the region CD, it cannot get to the region AB and vice versa.

Once again, quantum mechanics changes all these. We now have to associate a wave with a particle of energy E. Such a wave will look somewhat like the one shown in Fig. 3.17. In the region CD and AB, it will look roughly like a wave with an oscillatory behaviour but in the intermediate region it will decay rapidly from C to B. In physical terms, this means that while the particle is most likely to be found in the region CD, there is a very small but nonzero chance of it being found outside the potential barrier in the region BA. This is indicated by the presence of nonzero wave amplitude in the region BA. So, if you put a particle of energy E in the region CD, it can "tunnel through" the potential barrier and escape to the outside world! This is impossible in classical physics but is allowed in quantum theory.

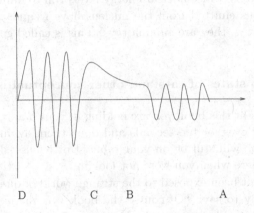

Fig. 3.17 The amplitude relevant to a tunnelling configuration. The particle has the usual wave like amplitude in the two regions CD and AB. But in the classically forbidden region the amplitude of the wave decays rapidly in going from C to B. While this makes the chance for the particle to go from C to B very small it remains nonzero in quantum theory.

The alpha decay of the nuclei proceeds along such a tunnelling process. The alpha particle inside the nucleus is similar to a particle of energy E confined in the region CD in Fig. 3.16. Every once in a while one such particle can tunnel through the potential well and thus escape out of the nuclei. Given the mathematical apparatus of quantum mechanics, it is possible to compute the rate at which an alpha particle of energy E will escape from a nucleus. The agreement between such a theoretical calculation and the observations was one of the early successes of quantum mechanics. It illustrated once again the practical application of a purely quantum mechanical process — the tunnelling through a potential barrier.

To complete the picture, it should be mentioned that the nuclei can also decay through two other processes called beta decay and gamma decay. Early experiments suggested that, in β-decay, the nuclei emitted electrons (in contrast to alpha particles) essentially converting the neutrons in the nuclei into protons. This was even more bizarre because there were no electrons in the nuclei to begin with! We will see later in Chapter 4 that this process required a much more sophisticated explanation rooted in relativistic quantum mechanics.

The nuclei of the atoms can also exist in different energy levels. When the nucleus makes a transition between two such levels, it can emit very high energy photons. Since nuclear energy levels run to millions of electron volts, the photons emitted from the nucleus have frequencies upwards of 2×10^{20} Hz; that is, they are gamma rays. This is called gamma decay.

3.7 Quantum states of cats and other conceptual issues

Take a good look at this book you are holding in your hand. Put it on your table, close your eyes for ten seconds and open them again. The chances are that the book will still be on your table. But let us ask the question: Was the book there when you were not looking?

If you have not been exposed to the strange world of quantum mechanics, you are likely to say: "Of course the book was there; the behaviour of the book does not depend on whether I am observing it or not." That, precisely is the point. Do physical systems behave differently when they are observed and when they are not? As we have seen in the case of two-slit experiment, quantum mechanics tells you that an electron behaves differently when you observe it compared to when you do not!

To understand the strangeness which lurks in the quantum theory, one has to take a careful look at the quantum mechanical description of a physical system and compare it with the classical description. Consider, once again, the motion of a particle from some location A to another location B. Suppose a ball is thrown from the location A at some instant of time and reaches the location B ten seconds later. If the forces which were acting on the ball are known, then one can compute the exact path taken by the particle in going from A to B. What is more, one will be able to calculate the position and speed of the particle at any instant while it was in flight. The situation, as we saw earlier, is very different in quantum theory. The uncertainty principle prevents us from specifying both position and speed of the particle simultaneously. Thus if we know the position at any given instant of time precisely then we will be completely ignorant of its speed at that instant. Hence we will not know where the particle will be at the next instant. Quite clearly this rules out the possibility of describing the motion in terms of a given path. Instead of a single path connecting A and B we now have to deal with *all possible paths* connecting the two points. A quantum mechanical particle which starts at A and reaches B could have followed any one of these paths connecting A and B. There is some chance for the particle to have followed any one given path among all possible paths.

This description in terms of probabilities is central to quantum mechanics and arises in all other contexts as well. Consider, as a second example, the spin state of an electron emitted by an electron gun. Quantum mechanics tells us that the spin of an electron along any specified direction, say the vertical direction, can take the values $+(1/2)$ or $-(1/2)$. In general, we have no way of knowing in which of these two states the electron will be when it is emitted by an electron gun. All that one can say is that it has some chance to be "spin-up" and some chance to be "spin-down". Let us denote by the symbol $|\text{up} >$ the quantum state in which the electron spin is $+(1/2)$ and by $|\text{down} >$ the state in which the electron state is $-(1/2)$. These two states are allowed in the mathematical description of quantum mechanics. But these are by no means the only states which are allowed in quantum theory! The equations of quantum theory allow several combinations of these two basic quantum states like, say, the state $(|\text{up} > +|\text{down} >)$. (This is essentially the result of the superposition principle which we mentioned in Sec. 3.4.) In such a general state, one cannot attribute a definite spin to the electron.

Suppose we now devise an experiment to measure the spin of the electron. The result of the measurement will always give the value $+(1/2)$ if the electron is in the state $|\text{up} >$ and will give a value $-(1/2)$ if it is in the state $|\text{down} >$. What happens if it is in a combined state like $(|\text{up} > +|\text{down} >)$? Then we will sometimes get the result $+(1/2)$ and sometimes get $-(1/2)$. But what is really surprising is the following fact. Suppose we measure the spin of an electron emitted by an electron gun and find that it is in $+(1/2)$ state. If we now measure its spin again we will get *only* $+(1/2)$ and never $-(1/2)$. In other words, the first act of measurement has somehow changed the complicated original state of the electron to the state $|\text{up} >$! Thus the act of observation or measurement has disturbed the quantum system in a very specific way.

The situation is similar for other quantum measurements. Suppose we try to nail down the path taken by an electron. For example, we could set up a complicated series of devices in order to actually determine whether the electron went through a particular route or not. If we do that, then the electron will start behaving as suggested by our intuition; its wave nature disappears. In other words, the behaviour of the electron depends on whether we are observing it or not. Even though the above result may appear strange and counter-intuitive, it has been routinely verified in numerous quantum mechanical experiments. These experimental results show that the evolution of the system does depend on whether it is being observed or not.

This fact raises a deep philosophical problem. How can one make sense of a separation between the system being observed and the detecting apparatus? At a fundamental level, both are subject to the same laws of (quantum) physics. Several difficulties arise when one artificially separates the physical system into a part which is being observed and the instrument which is performing the observation. These difficulties are brought out vividly by the example, usually called the "Schrodinger's cat".

One version of this example runs as follows: Think of a completely sealed room, inside which we place a cat and a device containing a poisonous gas. The device is set to be triggered by a purely quantum mechanical event like the decay of a radioactive atom. When the decay occurs, the device is triggered releasing the poisonous gas which will instantaneously kill the cat. We set up this system at some given time and ask — at some later instant of time — the question: "Is the cat now dead or alive?" Clearly this is equivalent to asking whether the radioactive atom has 'decayed' or not, which is a purely quantum mechanical question. Let $|0>$ and $|1>$ denote the quantum states in which the atom has 'decayed' and 'not-decayed' respectively. These are similar to the states $|up>$ and $|down>$ for the electron spin discussed earlier. But just as in the case of the spin, the system *need not* exist in either of these simple states. In fact, the laws of quantum mechanics tells us that the quantum state of the system in our experiment will be a mixed combination of $|0>$ and $|1>$. In such a mixed state there is some chance for the atom to have decayed and some chance for the atom not to have decayed. *The same goes to the cat too!* It is neither dead nor alive but should exist in a mixed state. Such a state may be called a state obtained by "linear superposition" of two intuitively obvious states — in this case the "dead state" and the "live state".

But have you even seen a cat in such a state? No! Macroscopic systems are almost never seen in such weird superposed states even though such states are perfectly legitimate in quantum theory. To drive the point home, consider another example. Quantum mechanics allows a billiard ball to be at a position A (let us call this state $|A>$) or at another location B (state $|B>$). But it also allows the billiard ball to exist in a state in which it is neither here nor there; a state which could be mathematically characterized as, say, $|A> +|B>$. The trouble is such states are almost never seen in the real world for macroscopic objects even though there is nothing in quantum mechanics that prevents their occurrence. This fact, clearly, requires an explanation.

The conventional interpretation of quantum mechanics does not offer much of an explanation for this phenomenon. The interpretation relies on a hypothesis called "collapse of the quantum state." The central idea is the following. It is assumed that the act of measurement makes the system "jump into" one of the natural states. For example, the billiard ball could be in the state $|A> +|B>$ as long as it is not observed. The moment we perform a measurement to ascertain the position of the ball the state will change into either $|A>$ or $|B>$. In other words, we have a fifty-fifty chance of finding the billiard ball at position A or position B and no observation will catch it in the mixed state. The same goes for the cat in the sealed box. When we open the box and peep inside, that act of observation will make the state become either "dead" or "live."

Such an interpretation depends crucially on the existence of an external apparatus which can be used to make the measurements. In fact such an apparatus must be macroscopic and obey the classical laws of physics. The ultimate classical apparatus performing the observations is, of course, the human brain which has to come into play in any sequence of complicated recording processes. This artificial division of a physical system into a classical observer and a quantum mechanical subsystem has been the most unsatisfactory feature of the above interpretation. What is worse, this interpretation has some mysterious overtones regarding the nature of human brain and the role of consciousness. Such issues have increased the confusion which surrounds this matter. (Interestingly enough, most working physicists did not worry too much about these "philosophical" issues. The reason was quite simple. Though these issues are important at a fundamental level, they did not affect the results of any computations. Practical applications of quantum mechanics to physical systems did not suffer due to the lack of answers to the above questions in most situations.) But if we really need to make sense out of a quantum universe, it is important to reformulate quantum theory without the notion of external observers.

Remarkably enough, it seems that this task can be achieved and the key idea is something called the "decoherence". Decoherence is a physical process which can make *large*, macroscopic, systems to behave classically — that is, behave in a way we expect them to behave. What is more, this process is directly tied to the fact that classical systems are "large". In fact, the degree of classical behaviour exhibited by a system will be directly related to the size of the system.

To see how this process works, let us go back to our discussion of the Schrodinger's cat. Why is it that cats are never seen in a state which is a superposition of "live" and "dead" states? Or, why do we never see billiard balls and tables at states with no definite location in space? To understand this feature we have to look deeper into what we mean by a state. The state of a truly quantum mechanical object — like an electron in a hydrogen atom, say — can be specified by very few parameters. In fact, one needs only three integers to specify the state of an electron in a hydrogen atom. If we have now a hydrogen molecule we would require more parameters: three each to specify the states of the electron in each of the atom and a few more to describe the mutual interaction between the atoms. In general, the larger the system the larger will be the number of parameters needed to specify the state of the system. If a system is made of N particles, then one typically requires $3N$ numbers to specify the state of the system. (The actual result is somewhat more complicated but this simplified picture is adequate for our purpose.) A cat weighing one kilogram will have about 10^{26} atoms in its body; quantum mechanics tells you that if you want to specify the state of the cat completely, one has to specify about 10^{26} independent parameters! Such a description will be purely quantum mechanical and will tell you everything about the cat. But we certainly don't do anything so crazy. When one thinks of a cat sitting in the corner of a room, we specify far fewer parameters than is needed for the complete specification of the *quantum* system. In other words, there are several possible quantum states of a cat which will be acceptable to us as a fair description of the situation: "there is a cat sitting at the corner of this room."

The fact that we are working with far less information than is needed to specify a quantum mechanical system changes the results drastically. Suppose we first develop a full quantum theory of a cat and then ignore all the unobserved degrees of freedom. This can be done mathematically by using a device called "density matrix". Such a computation shows that the effect of unobserved degrees of freedom is to make the system behave as though it is classical! In other words, macroscopic systems are classical for the very reason that they are macroscopic and that we actually deal with far fewer parameters than is required to describe the system fully.

This beautiful result shows that there is no *fundamental* difference between a classical system and a quantum mechanical system. A system appears to behave more and more classically as we start ignoring a large number of internal parameters of the system. In fact, this theory suggests that if we devise an experiment to measure all the parameters specifying a cat, it will behave as quantum mechanically as an electron.

It is conventional to call the unobserved degrees of freedom as the "environment." This idea has a wider applicability. There are several situations in which a system we are studying cannot be considered to be really isolated but must have a weak coupling to its surroundings. In such a case, one cannot provide a complete description of the systems without taking into account what happens to the environment. But very often we cannot observe and specify the quantum state of the environment completely and hence have to ignore several parameters which may have a bearing on the description of the system. In the case of the cat, we divided the full set of parameters into two parts: those which we observe and those which we don't. The unobserved set of parameters can be thought of as providing the 'environment' in this case.

Chapter 4

Particles, particles everywhere

4.1　Adding relativity to quantum theory

The description of quantum world in the last chapter was based on non-relativistic mechanics, in the sense that we never talked about the speed of light, the relative nature of time shown by clocks, etc. while describing the quantum mechanics. But we know that the real world is relativistic and hence the description of quantum mechanics which we provided in the last chapter must be modified by incorporating the principles of relativity. You might think at first that this should be fairly straightforward and should lead to differences somewhat like the differences between relativistic mechanics and Newtonian mechanics which we described in Chapter 2. Well, it turns out that life is not quite so simple.

To begin with, let us distinguish between situations which require general relativity — typically those involving strong gravitational fields — and those which only require special relativity. The former case turns out to be so intricate conceptually that we still have not quite managed to come up with a viable description. This is the so called problem of quantum gravity that we will describe in Chapter 6. But if we ignore the existence of gravitational fields for the moment, then we only need to put together the principles of special relativity and quantum theory. This has been achieved (or so we believe!) but, as you will see, only after the very foundation of quantum mechanics we described in the last chapter is shaken and redone. That is the theme we will develop in this chapter.

Why should putting together the principles of special relativity and quantum theory be so non-trivial? Briefly stated, it has to do with the fact that the concept of a particle becomes fuzzy when special relativity is brought in — which, in turn is related to the fact that energy and mass are interconvertible. Consider the collision between two particles of sufficiently high energies. In non-relativistic mechanics, the two particles will remain two particles after the collision but will, of course, exchange energy and momentum. In special relativity, however, another possibility exists. Part of the kinetic energy of the two particles involved in the collision could be converted into mass (your $E = mc^2$ at work again) and you can end up producing a few more particles in the process. In fact, much of experimental high energy physics is done by precisely this procedure in which you produce new particles by colliding a given set of particles with sufficiently high energy. You can immediately see that this is going to give you trouble when we bring in quantum theory. In the description of quantum mechanics we associate a quantum state or a wave amplitude *with a given particle*. Such a description cannot handle sudden appearance and disappearance of particles which is what special relativity ushers in. Roughly speaking, quantum mechanics is a theory for fixed number of particles but relativity demands that we introduce a mechanism for creation and annihilation of particles.

This may not sound too drastic but it actually is. To see that, let us explore the whole idea from a different and more elegant angle. We saw in Chapter 3 that one key feature of quantum mechanics was described by Fig. 3.5 which explains how to compute the chance for a particle at a given position x_1 at time t_1 to end up at another position x_2 at time t_2. Quantum mechanics tells you that you should draw all the paths which connect the two points, associate a particular chance for a particle to go along any given path and add them all up. But if you look at Fig. 3.5 carefully, you will realize that we have actually not drawn *all* possible paths which connect the two points. There is something very special about the kind of paths which are drawn and used in Fig. 3.5 which is made clear in Fig. 4.1. The vertical line in Fig. 4.1 often intersects any given path at more than one point. (It is illustrated by singling out one path by a thick line.) That is, a given path could represent a particle which was at the same spatial location (denoted by the coordinate x_a) at more than one instant of time. There is nothing wrong in having such a path. After all, if you throw a stone vertically up — such that it reaches a maximum height and falls back to Earth — it will be at some given height above the ground once while going up and also while coming down. We see these kinds of paths all the time.

But now consider a horizontal line in Fig. 4.1 which denotes a given instant in time. You notice that none of the paths in Fig. 4.1 cut a horizontal line at more than one point! So clearly, the paths which are used in Fig. 4.1

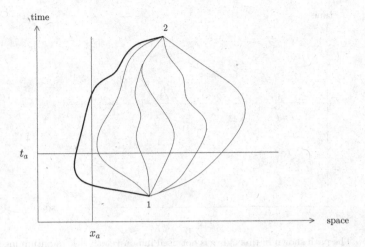

Fig. 4.1 In non-relativistic quantum mechanics the amplitude for a particle to go from one point to another is obtained by summing over the amplitudes for different paths. But these paths, like the ones shown here, deal with space and time differently. Horizontal lines (represssenting a constant time) cut each of these paths (like the one shown by the thick line for illustration) only once. But the vertical line (representing a given location in space) can cut a given path several times. This difference arises because we expect the particle to be at a given location at any time but nothing prevents a particle from being at the same spatial location at different times.

are rather special: We ensure that a given path never cuts a horizontal line at more than one point.

To make clear what we mean by this, take a look at Fig. 4.2 which has a path that violates the above restriction. This path cuts a horizontal axis in 3 points. Why is it that we never considered such paths in Fig. 3.5? A horizontal axis represents a given instant in time and we tacitly assumed that the particle is at some given location at any given instant in time. A path like the one in Fig. 4.2 would require us to conclude that our particle was at 3 different spatial locations, ①, ② and ③ simultaneously at a given instant in time; rather, at the instant of time represented by the horizontal line in Fig. 4.2, we had three particles in existence rather than a single particle. Though quantum mechanics is weird and allows a particle to explore all possible paths between two events, we limited its weirdness by saying that at any given time the particle could have been in only one place.

It is now obvious that we treated space and time very differently in the above discussion. A particle could go back and forth in space so that it

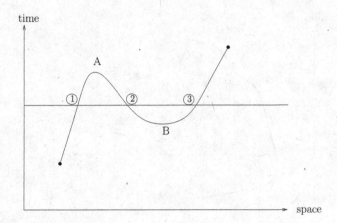

Fig. 4.2   The path shown in this figure is not used in non-relativistic quantum mechanics in computing the amplitude for propagation. This is because the path here represents a trajectory in which a particle can be at three different locations at a given time. It is, however, necessary to take such paths into account while formulating a relativistic quantum theory.

could cut a vertical line several times. But a particle cannot go "backward in time" (from A to B in Fig. 4.2) so that it cannot intersect a horizontal line more than once. This is perfectly fine in a non-relativistic description in which space is space and time is time and they don't mix. Time is absolute and we can always treat time as quite distinct from space.

It is exactly at this point when Einstein steps in and spoils the party. Special relativity requires a greater level of symmetry between space and time if all observers in different inertial frames have to describe a phenomenon in an identical fashion. In special relativity the proper time, shown by the clock carried by a particle, plays the role of the absolute time of Newtonian mechanics. As a particle moves along a trajectory, it has to always move forward in *its own proper time,* i.e., in the time shown in a clock carried by the particle. But there is no restriction that it should always move forward with respect to the time t shown by some other clock which you decide to use for labelling the coordinates. Such a clock time has only the same status as the spatial coordinates in non-relativistic physics.

This result is a lot easier to state using symbols: In non-relativistic physics a path is specified by giving three functions $x(t), y(t), z(t)$ which gives the spatial coordinates of the particle at any given absolute time t. In special relativity there is no absolute time. So the paths *in spacetime* is given by specifying 4 functions $t(\tau), x(\tau), y(\tau), z(\tau)$ which gives you the spatial locations *as well as* the coordinate time corresponding to a proper time τ shown by the clock moving with the particle. The particle cannot go backward in the t in non-relativistic physics but can go forwards or backwards in x, y, z. Similarly, a particle cannot go backward in the proper time τ in relativistic physics but it can go backwards or forwards in x, y, z *as well as* in t. The last bit is the sting in the tail which relativity brings in. It means that paths like the one in Fig. 4.2 must be allowed in relativistic quantum mechanics.

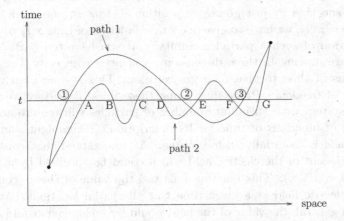

Fig. 4.3 Two representative paths that need to be considered in relativistic quantum theory. The first path cuts the horizontal line (represting a given moment of time) in three places while the second path cuts it at seven places. We can draw more complicated paths which could cut it at arbitrarily large number of points. This will require considering several particles being simultaneously present at different spatial locations.

That requires a new revolution. We are now forced to admit that when we combine quantum theory with special relativity, there is a chance for paths like the ones shown in Fig. 4.3 can occur with some chance. The path 1 in this figure cuts the horizontal axis at some time t at three different places marked 1, 2, and 3 while the path 2 in the figure cuts the horizontal line at 7 points $A, B....G$. This means that a single particle should appear as 3 different particles located at 1, 2 and 3 when it follows the path 1 and it will appear as 7 different particles when it follows the path 2. Of course, you can draw paths which wiggle a lot more; in fact, a path can wiggle an arbitrarily large number of times leading to the existence of an arbitrarily large number of particles at any given instant in time! Since relativity demands that we take into account these paths, we need to do two things. First, we need a description which can handle an infinite number of particles at any given instant in time. Second, we need a physical interpretation of what the wiggly paths in Fig. 4.2 and Fig. 4.3 physically represent.

The first task leads to the concept of *quantum fields* which we 'shall now describe. This idea is closely related to that of degrees of freedom which we had occasion to discuss in a different context in Chapter 1. A Newtonian particle moving in one dimension, say along the x-axis, is said to have a single degree of freedom in the sense that its state at any given time can be specified by just giving its position — that is, one single number. More generally, we have to give one single function of time $x(t)$ to describe the motion of such a particle. Similarly, if you have ten particles all of which are moving in three dimensions, you need to specify $3 \times 10 = 30$ functions of time to give their trajectories. This system clearly has 30 degrees of freedom. Proceeding along these lines, it is easy to conclude that a system with an infinite number of particles will require an infinite number of functions of time for its description. That might sound scary but *a field* is essentially such a system. To understand this, consider the $x-$component of the electric field which could be specified by a function $E_x(t, \mathbf{x}) = f(t, \mathbf{x})$. This function tells you the value of the $x-$component of the electric field at a given time t at all spatial locations. At a later time, in general, the value of the field would have changed at all locations. Such a description is equivalent to us specifying how the field behaves as a function of time at every location in space. If you take a set of points in space, $\mathbf{x}_1, \mathbf{x}_2, \mathbf{x}_3 ....$, and specify how the electric field changes with time in each of these locations, then we will have substantially the same description. Of course, you will need to know the behaviour of the field at *every* location in space. Since there are infinite number of points in space, you need to give an infinite number of functions of time to describe a field. It is in this sense — which can be made mathematically more precise by working in what is known as Fourier space — that a field is equivalent to a system with an infinite number of degrees of freedom. So the wiggly paths in Fig. 4.3 — which are mandatory when we combine relativity with quantum theory — require describing our system with an infinite number of degrees of freedom and the field provides a convenient way of doing this. Hence relativistic quantum theory is often called quantum field theory, which is a quantum description of a field.

In such a description we associate a field with every particle. There is a field associated with the photon (which is essentially the electromagnetic field, though the mathematical description requires working with certain potentials from which electric and magnetic fields can be obtained). There are also fields associated with an electron, with a proton, with a neutron etc. Just as photons emerge as a quanta of the electromagnetic field the electrons, for example, will emerge as the quanta of the electron field. The precise nature of the field that needs to be associated with a particle depends, among other things, on the spin of the particle. For example, a spin half particle like an electron (which, as we know, is a fermion) is called a spinor in mathematical jargon; the electric and magnetic fields associated with a photon are vector fields which is in sharp contrast with the spinor fields describing an electron. This difference lies at the basis of the historical fact that it was the wave nature of photons and the particle nature of electrons which became apparent initially. Understanding the particle nature of the electromagnetic radiation and the wave nature of the electron required considerable effort by the researchers. We now know that to every particle we can associate a field and, in fact, particles can be thought of as excitations of the underlying field. (This will become apparent if you have heard of certain particles called phonons. These are just the quanta of what we would normally call a sound wave propagating in a media. It has nothing to do with relativity as such but the same mathematical description applies. Condensed matter physics is full of such particles which are excitations of wave fields of different kinds.)

Let us now consider the second question, viz., the task of providing a physical interpretation for the wiggly paths. This leads to *the* most important result which arises when we put special relativity and quantum mechanics together. Figure 4.4 represents a prototype of a troublesome path which cuts the horizontal line at 3 different points. In this path, we have only one particle at all times prior to t_1; we have three particles at $t_1 < t < t_2$ and again one particle for $t > t_2$. One way of interpreting this path is to introduce a concept of an *antiparticle* for every particle and introduce a "rule" that a particle travelling backward in time is the same as an antiparticle travelling forward in time. Then we will interpret the path in Fig. 4.4 along the following lines. We start with a single particle going forward in time. At the event B, a particle–antiparticle pair is spontaneously created and the particle goes from B to Y. The antiparticle goes from B to A and annihilates the original particle which was going from X to A. So with one creation and one annihilation of pairs of particles at B and A, we have managed to transport a particle from X to Y. More importantly, both the particles and the antiparticles were travelling forward in time in this description. The antiparticle travelling from B to A (forward in time) is considered equivalent to the particle travelling from A to B (backward in time).

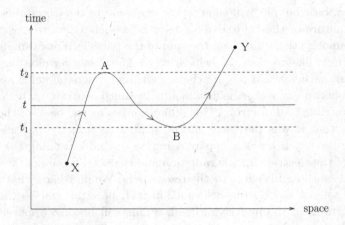

Fig. 4.4   The interpretation of paths like the one shown here requires the existence of an antiparticle. The antiparticle travelling from B to A (forward in time) is considered equivalent to the particle travelling from A to B (backward in time). The interpretation now starts with a single particle going forward in time. At the event B, a particle–antiparticle pair is spontaneously created and the particle goes from B to Y. The antiparticle goes from B to A and annihilates the original particle which was going from X to A. So with one creation and one annihilation of pairs of particles at B and A, we can transport a particle from X to Y.

This is utterly bizarre and takes lot of getting used to. So let us try to understand which are the new concepts we have smuggled in. To begin with, we must have an antiparticle to every particle if this description has to work. To an electron we need an antielectron, to a proton we need an antiproton, etc. Second, it must be possible to create and annihilate particles out of "nothing" if events at A and B have to be interpreted. This is possible because of the uncertainty principle. To create a pair of particles each of mass m, you need an energy $2mc^2$. The uncertainty relation between time and energy $\Delta E \Delta t \simeq \hbar$ says that you can violate the energy budget by an amount $\Delta E \approx 2mc^2$ for a short period of time $\Delta t \leq \hbar/2mc^2$ and no one needs to know. As long as the creation and annihilation from the vacuum takes place in such a short timescale, we can have processes like creation and annihilation at A and B occurring. Once again, this is a direct consequence of combining relativity ($E = mc^2$) with quantum theory $\Delta E \Delta t \simeq \hbar$.

The first issue, of the existence of an antiparticle for every particle, is more important and, of course, is of great practical importance. We have an immediate prediction from our crazy way of merging relativity and quantum theory: there must be, for example, an antielectron. This prediction came originally from P.A.M. Dirac, one of the greatest physicists of 20th century. With a little bit more mathematics (done in a different, but equivalent, way) one could predict the properties of the antielectron — which is more conventionally called a positron. A positron carries positive charge while the electron, as you know, is negatively charged. (The positron should not be confused with other positively charged particle we have come across, viz., the proton. The proton is nearly 2000 times heavier than the electron. What is more, as we shall see later in this chapter, the proton is made of subunits called quarks. The positron, on the other hand, has the same mass as the electron and — as far as we know — neither the positron nor the electron has any substructure in it.) The discovery of positron in the laboratory was one of the high points in the development of relativistic quantum theory and showed that, in spite of the apparent craziness, the ideas matched with what is seen in nature — which, as we have emphasized several times, is the true test of any physical theory. All other elementary particles have their own antiparticles. The proton, for example, has the antiparticle "antiproton" which is negatively charged.

Though the description of antiparticles might sound like science fiction, it is not. Positrons, antineutrons, antiprotons and many other kinds of antiparticles are routinely produced in the laboratory in the course of high energy interactions. The principle behind the production of any such particle is the same: in a sufficiently high energy collision one can convert part of the kinetic energy into matter; by suitable choice of the colliding particles one can produce the antiparticles mentioned above. The material world which we see around us is made of protons, neutrons and electrons, i.e., it is made of matter. One can imagine another (hypothetical) world made of antiprotons, antineutrons and positrons as the constituents of (anti)atom. For reasons which are not very clear, there exists a deep asymmetry between matter and antimatter in the world around us. Simply stated, we see a lot of matter, but virtually no antimatter. We shall see in a later chapter that this fact can have important cosmological implications.

4.2 The particle zoo

The two features introduced by relativity into the quantum mechanical description of nature — the existence of antiparticles and the possibility of creation and annihilation of particles — have far reaching implications. We shall now describe several of these consequences.

The existence of antiparticles has, at one stroke, doubled the total number of particles we have to deal with. One might have thought originally that all matter is built out of protons, neutrons and electrons and in that sense they are the building blocks of matter. But we now know that the antiparticles corresponding to these three particles (antiproton, antineutron and positron — the last one being the antiparticle of electron) must exist in nature. More importantly, our idea of certain particles being more "basic" or "elementary" needs to be carefully reviewed in the light of the possibility of creating and annihilating the particles.

For example, it was soon realized that there are very many different kinds of particles present in nature other than the proton, neutron and electron. The evidence for these particles came from different sources and in particular from cosmic rays. Cosmic rays are the name given to streams of high energy particles reaching us from the interstellar space of our galaxy. (These are extensively investigated but their primary source is still a mystery.) The cosmic ray which reaches Earth from outer space consists mostly of protons. When these high energy protons collide with the nuclei of atoms in the atmosphere, they give rise to all kinds of new particles which are not naturally seen in the laboratory. The study of cosmic rays was one of the key research areas from which information about a host of particles was obtained before the construction of large particle accelerators.

A typical process which is seen in a cosmic ray event involving several unusual particles is shown in Fig. 4.5. It begins with a particle called pion (π^-) which is spinless and carries negative electric charge. When a cosmic ray pion hits a proton in the nuclei of a molecule in the atmosphere, two new particles called Kaon (K^0) and Lambda (Λ^0) are produced. These two particles, in turn, can produce through their decay other particles like, for example, a positively and a negatively charged pion or a pion and a proton. This kind of an interaction illustrates that our naive ideas as to which particle is "made of" which particle simply will not work. What is involved is the transformation of energy into material particles and vice versa and one needs to have a more sophisticated procedure for describing the particles and their interactions.

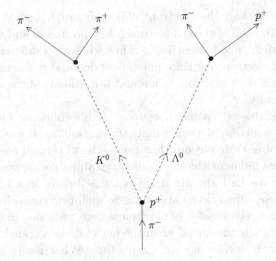

Fig. 4.5 One example of particle production in the atmosphere due to collision of a high energy cosmic ray with protons in the atmosphere. The study of events like this quickly led to a proliferation of particles and the naive interpretation of certain particles as 'elementary' fell by the way side.

This, in fact, raises several basic questions. If the collision of two protons, say, can generate new particles, should one imagine that the protons are actually "made up" of these particles? If so, what will happen when we collide the new set of particles obtained in this process? In other words, do we know what are the most fundamental units out of which all matter is made of?

Particle physicists have made significant progress in answering the above questions. There is considerable evidence to suggest that all matter is made of two basic kinds of particles called "quarks" and "leptons" with no evidence for any further substructure. The forces between these particles are mediated by another set of particles called "vector bosons". We will now look at these particles more closely.

4.2.1 *Leptons and quarks*

The simplest of these three categories is the leptons. There are six different leptons which exist in nature, of which the electron is the most familiar one. There are two more particles called "muon" and "tau" lepton which are identical to the electron, except in their mass. The muon is about

200 times heavier than the electron while the tau lepton is about 3000 times heavier. To each of these 3 leptons (electron, muon and tau) one can associate a particle called "neutrino". Thus we have 3 different species of neutrinos called electron-neutrino, muon-neutrino and tau-neutrino. These 3 neutrinos, along with electron, muon and tau constitute the six members of the lepton family.

The neutrino deserves some discussion in this context. The physicists were lead to the neutrino in a somewhat strange fashion. It was known from the early days of nuclear physics that the nuclei of certain elements could decay along three different channels called the alpha decay, beta decay and gamma decay. We had already discussed this briefly in Chapter 3 (see Sec. 3.6). Of these, alpha decay arises due to quantum tunnelling while the gamma decay involved emission of electromagnetic radiation in the form of a gamma ray photon by the nuclei when the nuclei make a transition between two internal energy levels and emit a photon. What really troubled the physicists was the phenomena called beta decay in which *electrons* streamed out of the nucleus. This was mysterious for several reasons. To begin with, it was not clear how electrons could have existed inside the nucleus in the first place. If you confine the position of the electron to a size smaller than that of a nucleus, then the uncertainty principle would require attributing to it a fairly high amount of momentum and corresponding amount of energy. As a result, it cannot be confined within the nucleus. Second, careful measurement showed that the energy carried away by the electrons emitted in the beta decay was not always the same and, on the average, it was less than the energy lost by the nucleus. Either some energy is being sneaked away by an unknown process or the energy conservation is being violated. There was also a corresponding difficulty about the conservation of angular momentum when we take into account the spin of the nuclei before and after the beta decay and the spin of the emitted electron.

The resolution of these problems was first suggested by W. Pauli. He postulated a new particle now called neutrino which was being emitted in the beta decay along with the electrons. Soon E. Fermi developed a theory in which a neutron in the nuclei could transform itself into an electron, a proton and an antineutrino. The proton stays within the nucleus while the electron and the antineutrino stream out of the nucleus. Since the neutrino can carry energy and angular momentum, it was possible to satisfy the conservation laws in the beta decay. Most physicists accepted the reality of the neutrino after some initial debate though it was not directly observed until 1956. The neutrinos have no charge and undergo only the weak interaction.

As a result, a neutrino with the energy of a few million electron volts has a better than even chance of passing through a solid wall nearly 10 light years thick! This explains why the neutrino is often called a ghostly particle.

The real genius of Fermi in developing the theory for beta decay is probably in introducing the concept of decay of elementary particles. The key to his theory lies in the reaction, $n \rightarrow p + e + \bar{\nu}$. Here is a process in which one particle (neutron) spontaneously disappears and three particles (proton, electron and antineutrino) appear in its place. Normally one would have thought that such a reaction implies that the neutron is "made of" the proton, electron and antineutrino. This, of course, is simply not true. As we have seen before, creation and annihilation of particles is a rather routine matter in the effervescent world of relativistic quantum mechanics. In the theoretical model developed by Fermi, the neutron gets annihilated and in its place a proton, electron and antineutrino are created. So it will be completely wrong to think of a neutron as being made of proton, electron and an antineutrino.

All these six leptons — electron, muon, tau and the corresponding neutrinos — exist in nature and have been produced in laboratories during high energy collisions. Of course only one of these six particles, viz. the electron, exists in the atom under normal circumstances. This does not mean in any way that the other particles are not "real". They are all members of the lepton family and enjoy equal status in particle physics.

A more dramatic way of presenting this result is to say that the leptons come in three flavours with each of the flavours being identified with the electron, muon or tauon. Correspondingly, the neutrinos also come in three flavours and are associated with these leptons. The concept of a flavour is not just a window decoration. To see this, consider the following reaction, $\mu \rightarrow e + \gamma$, which denotes a muon decaying into an electron emitting the excess mass energy in the form of a photon. Such a reaction was never observed in nature. Usually, physicists try to understand the existence or nonexistence of reactions in terms of conservation laws. One would expect a reaction to conserve mass-energy, angular momentum and — of course — the electric charge. The above reaction can conserve all these and yet seems to be forbidden in nature. However, if we attribute the concept of a flavour to the leptons, which is also conserved in the reactions, then one can understand why this reaction cannot occur. The flavours on the left side and right side of the reaction are different preventing this reaction from occurring.

It is, however, possible for the muon to decay into an electron through

Table 4.1 Elementary Particles

Particle	Name	Description
Leptons	electron	mass = 0.5 MeV (stable)
	muon	mass = 107.6 MeV (unstable)
	tau	mass = 1784.2 MeV (unstable)
	electron neutrino ⎫	
	muon neutrino ⎬	mass uncertain
	tau neutrino ⎭	
Quarks	up quark ⎫	constituents of
	down quark ⎭	protons and neutrons
	top quark	
	bottom quark	Not seen as
	charm quark	free particles
	strange quark	
Vector bosons	photon	[mediates electromagnetism]
	W-boson	[mediates weak force]
	Z-boson	[mediates weak force]
	gluons	[mediates strong force]
	(gravitons)	[mediates gravity]

the reaction $\mu \rightarrow e + \nu_\mu + \bar{\nu}_e$ where ν_μ is the muon neutrino (which has the same flavour as the muon) and $\bar{\nu}_e$ is the antineutrino associated with the electron (which has the flavour opposite to that of the electron). You can easily convince yourself that in this reaction the flavour is indeed conserved. In that sense, one can think of flavour as another attribute of particles like charge or spin which is conserved in the reaction.

Let us next consider the quarks. There are also 6 different kinds of quarks. They are denoted by the symbols "u", "d", "c", "s", "t" and "b" which stand for "up", "down", "charm", "strange", "top" and "bottom" quarks. Particles like protons, neutrons (and a vast majority of other particles seen in high energy collisions) are made of different combinations of these quarks. For example, a proton is made of two up-quarks and one down-quark while a neutron is made of two down-quarks and one up-quark. Since all matter has protons and neutrons we might say that — indirectly — all matter contains up-quarks and down-quarks. The other four types of quarks are not present in the normal state of matter. However, they are needed as constituents of several other elementary particles which are produced in high energy collisions. Once again, just like tau-leptons or neutrinos, these quarks are also needed to explain the high energy interactions.

There is, however, one key difference between leptons and quarks. Unlike leptons, the quarks are never seen as free particles in laboratory experiments. The fact that they exist inside a proton can be ascertained only by hitting the proton with very high energy particles which can penetrate into the proton. In such experiments — which probe the proton deeply — one could see that quarks exist as sub-units inside the proton.

The third category of particles is called "vector bosons". These particles play the important role of mediating interactions between other particles. According to quantum theory, all forces arise due to the exchange of some fundamental particles. Consider, for example, the electrical repulsion between two electrons. In quantum theory it is described by a process shown schematically in the top left frame of Fig. 4.6. (We will discuss this concept in detail in the next section.) We assume that the first electron emits a particle (in this case, called a photon) which is absorbed by the second electron. It is this transfer of a photon which actually produces the electrical repulsion between the two particles. In fact, the simplest of the vector bosons is a photon, which is the quantum of the electromagnetic radiation. All electromagnetic interactions between charged particles can be thought of as arising due to exchange of photons. Other forces of interaction are mediated by some other vector bosons. The weak nuclear force between quarks or leptons is mediated through the exchange of certain particles called "W" and "Z" bosons and the strong nuclear force between the quarks is mediated by a set of 8 particles called "gluons" (see Fig. 4.6). The gravitational interaction is thought to arise out of certain particles called gravitons. Of all these particles, photons are the most familiar and we shall discuss them at length in the next section; the W and Z bosons have actually been produced in the laboratory while the gluons are seen in some of the particle interactions indirectly. Thus the existence of these particles is well founded. The gravitons have not been seen directly and should be thought of as a theoretical speculation at present especially since we do not yet have sensible model for quantum gravity; see Sec. 6.1.

Thus, according to our current understanding, we can interpret all forces of nature using the picture of quarks, leptons and vector bosons. Of these, quarks and leptons act as basic building blocks of matter while the vector bosons mediate the interaction between the particles.

It must be remembered that the above picture is based on experimental results which are currently available in the laboratory and on certain

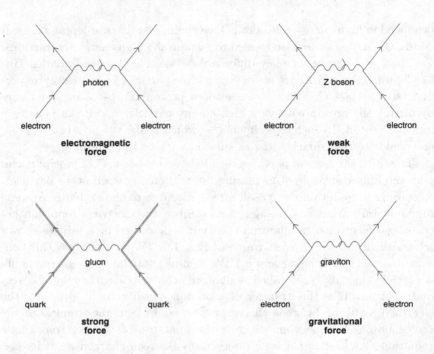

Fig. 4.6 According to quantum theory, all forces of interaction are caused by exchange of certain particles called vector bosons. For example, the electromagnetic force between two charged particles, say electrons, is caused by the exchange of a particle called the photon which is the simplest of the vector bosons. The strong force between two quarks, say, is caused by the exchange of a vector boson called the gluon. The weak force is similarly caused by the exchange of three vector bosons called the Z-boson, W^+ boson and W^- boson. It is believed that gravitational force is similarly due to the exchange of a particle called graviton.

theoretical models which are developed using them as inputs. At present, laboratories can only probe phenomena taking place at energy scales less than about 1000 GeV. Consequently we are in no position to detect a particle in the laboratory if its mass is considerably higher than 1000 GeV. (For comparison, remember that the proton has a mass of about 1 GeV.) The behaviour of particles at higher energies is based on theoretical conjectures. It is often possible to come up with *different* theoretical models at high energies which can explain all known experimental observations at low energies. But each of these models will postulate the existence of new kinds of particles which, because of their high mass, would not have been seen in the laboratories yet. For example, there exists a class of theoreti-

cal models called "supersymmetric models" which postulate the existence
of several new elementary particles in addition to the quarks, leptons and
vector bosons described above. So far, we have not seen any evidence in
the laboratory for the existence of these new elementary particles. Very
often, cosmological and astrophysical considerations could provide information about such particles — if they exist. We shall see in later chapters
that such models may have very important astrophysical and cosmological
consequences. Conversely, it will be possible to put constraints on such
models based on cosmological and astrophysical observations.

4.3   Quantum electrodynamics: A prototype model

The ideas presented in the previous sections lie at the basis of any theory which describes the interaction of particles taking into account both
the principles of quantum theory and relativity. Such theories are usually called quantum field theories. It turns out that constructing a viable
quantum field theory is no joke. For example, we still do not know how
to construct a viable quantum theory describing gravitational interaction!
We will now describe how quantum field theories are constructed and why
they sometimes work and sometimes don't.

   We will illustrate this by using the most successful of all quantum field
theories called *quantum electrodynamics* which describes a world made of
just electrons, positrons and photons. In such a model, one is essentially
interested in describing how the electromagnetic interaction of electrons can
be described. For example, we could let two electrons scatter off each other.
Classically, we will use the repulsive force between the two like charges to
describe it; the question is how we can do this taking into account the
quantum mechanical and relativistic effects.

   Figure 4.7 illustrates the a simple starting point for such a description.
This picture represents the scattering between two electrons due to, what
is usually called, "the exchange of one quanta". The solid lines in the figure
represents spacetime trajectories of the two electrons. At the event A the
first electron emits a photon which is absorbed at event B by the second
electron. The propagation of the photon in the spacetime is denoted by the
wavy line marked by the Greek symbol γ. The points A and B are called
vertices.

   The chance for this process to occur can now be written down as a
product of 3 factors. First, there is a chance that the electron will emit a

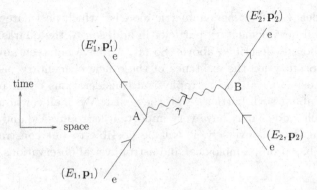

Fig. 4.7   The scattering of two electrons can be interpreted as due to exchange of a photon which transports energy and momentum between them. This diagram is easy to interpret and is the simplest process in quantum electrodynamics.

photon. Second, we need to take into account the probability amplitude for the photon to go from point A to point B. Finally, there is some chance for the second electron to absorb this photon. It is possible to give "rules" for each of these factors and multiplying them together, one can write down the total chance for this event to take place. The initial state of the two electrons can be given in terms of their initial energies and momenta which are taken to be $(E_1, \mathbf{p}_1)$ and $(E_2, \mathbf{p}_2)$. The corresponding final energies and momenta are denoted with a prime: $(E_1', \mathbf{p}_1')$ and $(E_2', \mathbf{p}_2')$. Once we compute the chance for such a process to take place, we can convert it into the probability for such a scattering to occur, which changes the momenta and energies of the two electrons. Of course, the classically specified force between the two electrons does exactly the same: i.e., it changes the momenta of the interacting particles. Thus the process in Fig. 4.7 essentially describes the force of repulsion between the two particles. We say that the electromagnetic force between charged particles is mediated by the exchange of photons, which are the quanta of the electromagnetic field.

There are several curious features which need to be noted about such a diagram. First is that we expect all the appropriate conservation laws to hold at every vertex. The incoming electron, outgoing electron and the photon at A, for example, should carry energy, momentum, angular momentum and charge such that they are conserved at A. Second, the initial and final electrons are real particles which are being scattered in the lab. Therefore, their energy and momentum must satisfy a specific relation which depends on the mass of the electron. It will turn out, however, that once we take care of this, the energy and momentum of the photon are completely specified and will no longer obey the relation $E = pc$! This is, however, fine because the photon in Fig. 4.7 is an intermediary and is not directly observed in the lab. It is called a virtual photon.

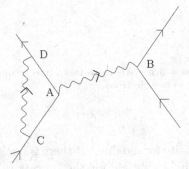

Fig. 4.8   This figure and the next show more complicated processes which need to be taken into account in studying the scattering of two electrons in quantum electrodynamics.

If that is all there is to it, then life would have been simple (and quite uninteresting) because the quantum field theory will essentially lead to results that are the same in classical scattering between the two electrons. But there is more to it — as can be seen from Figs. 4.8 and 4.9. These represent processes which can also take place in the interaction between two electrons in the sense that there is a nonzero chance for such processes to take place. Figure 4.8 represents the same situation as in Fig. 4.7 except that, in the former the electron also emits a photon at C and re-absorbs it at D. The chance for this process now has to be computed by including the factors which describe the process of emission at C, propagation of a photon from C to D and absorption at D. Figure 4.9 is a still more interesting example. Here the photon converts itself into an electron-positron pair at C; the electron and positron annihilate each other D to reproduce the photon. In computing the amplitude for two electrons to scatter off each other, we have to take into account the amplitudes for processes like the ones shown in Fig. 4.8 and Fig. 4.9. It is clear that you can draw several such diagrams of increasing complexity. For example, one could have added two electron positron loops or three or four etc. in a figure like Fig. 4.9. We need to take all these into account in a quantum field theoretic calculation.

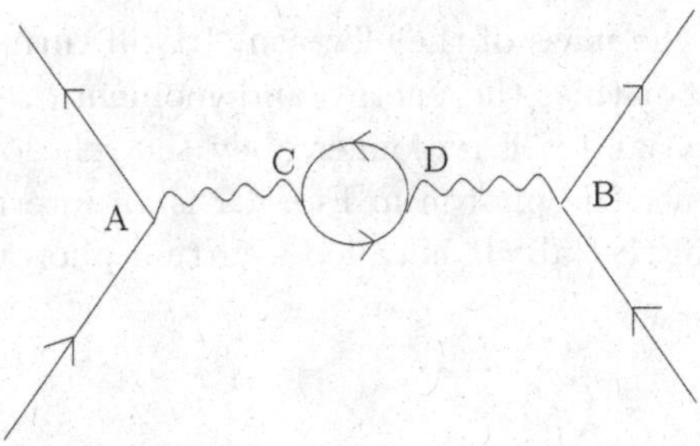

Fig. 4.9   Another process which needs to be taken into account in calculating the amplitude for the scattering of two electrons.

That might look quite formidable but there is one saving grace. We said that, in computing the amplitude for the process in Fig. 4.7, we need to multiply three different factors of which two correspond to the vertices A and B. It turns out that when you work out the mathematics, each vertex contributes a factor $\alpha = (e^2/\hbar c)$ where e is the charge of the electron. This factor is dimensionless and is called the fine structure constant. Its numerical value is about $1/137$; that is, it is approximately 0.01. The fact that the fine structure constant α is fairly small helps one in deciding which processes will dominate over which processes. If a diagram has two vertices, one will get a contribution of two factors of α from that diagram. This is what, for example, will happen in the case of Fig. 4.7. In contrast, Fig. 4.8 and Fig. 4.9 have four vertices in them; therefore, the amplitude for these processes will have four factors of α. More to the point, they will be a factor $\alpha^2 \approx 0.0001$ smaller compared to the process in Fig. 4.7. It therefore makes sense to compute our results as a series expansion in powers of α. We can compute the first few of them to get the result accurate to some level and ignore the remaining correction which will be smaller compared to what we have computed. Such a process is known as a perturbative calculation and this is the key to the success in studying electromagnetic interactions of electrons and positrons.

When this was attempted, people encountered a new, serious difficulty. To understand this, we need to realize that the factor α mentioned above is only part of the story in computing the amplitude for the process shown in, for example, Fig. 4.8 or Fig. 4.9. When the photon transforms itself into an electron-positron pair at C, it could do so with any momentum and energy for the virtual electron and positron. So the diagram in Fig. 4.9 represents one of the infinite number of diagrams we can draw each labelled by, say, the momentum of the virtual electron propagating between C and D. You also need to sum over all the possible momenta the virtual electron can have in order to compute the probability amplitude from this diagram. It turns out that, when you do this calculation, you get an infinitely large amplitude for this process! So the result could be written as a product of a factor involving the coupling constant α (which is, of course, a small number) multiplied by an infinity (which, of course, is the largest number you can imagine!). Since a finite number — however small — multiplied by infinity is still infinitely large, your calculation does not make sense. Each of the higher order processes like the ones in Fig. 4.8 and Fig. 4.9 lead to an infinite amplitude for the scattering of an electron by another electron which is simply absurd.

This absurdity is called the divergence problem in quantum theory. Divergence essentially means that you get an infinitely large answer for a quantity which should be finite and this is pretty much the worst thing that can happen in a physics calculation. It may be recalled that it is precisely this kind of trouble, with the radiation energy density inside a box in a classical calculation, that led to the quantum revolution in the first place. We now find that quantum field theory brings in infinities of its own which again needs to be tackled.

How do you handle this new problem? It turns out that there is a clever trick (usually called renormalization though the term regularization is a better one) by which you can tame the infinities and obtain sensible results in quantum electrodynamics. But this trick works only under very special circumstances — that is, only for particular class of theories. Considering its importance, we will describe the basic idea in spite of the fact that it is a bit technical.

The key idea is to note that the amplitude for different processes which we compute always depends on some parameters in the theory like, for example, the mass of the electron or its charge. (In a more complicated theory it could depend on more parameters.) These parameters need to be determined by performing different experiments in the laboratory which, in turn, will involve scattering processes like the ones in Fig. 4.7 to Fig. 4.9. For example, you can determine the charge of the particle by measuring the force acting on the particle (which is given in terms of the change in the momentum) when it is at a fixed distance from another charged particle. When such measurements are made, all the processes in the figures we have drawn — as well as many other more complicated figures which we have not drawn — contribute to it. We also know from direct observation that the observed mass, charge etc. of the electron are finite quantities. Let us assume, for a moment, that the electron has a bare mass m_{bare} and a bare charge e_{bare} which is the value of the mass and the charge in the absence of any interaction. (Obviously, it is a theoretical, unobservable, concept.) Further, suppose that the net effect of all the processes which lead to infinite contributions can be shown to be equivalent to replacing the bare mass and bare charge of the electron by two other parameters $m_{obs} = m_{bare}F_1$ and $e_{obs} = e_{bare}F_2$ where F_2 and F_1 are infinitely large numbers. We now declare that the new parameters m_{obs} and e_{obs} should be taken as equal to the observed mass and charge of the electron. In that case, we don't have to worry about the infinitely large factors F_1 and F_2 which are absorbed by "renormalizing" the mass and charge of the electron.

The astute reader will note that, in the relation $m_{obs} = m_{bare}F_1$ we can have a finite value for the left hand side with an infinite value for F_1 only if the bare mass m_{bare} is infinitesimally small. This makes the operation somewhat dubious but remarkably enough the idea works.

To demystify this a little bit, it is worth stressing that the basic fact — viz., interactions change the value of electron mass — is not as bizarre as it might seem. It is well known that the effective mass of an electron (as measured by the amount of force one needs to apply for it to acquire a certain acceleration) in a crystal lattice is different from its value outside the crystal lattice. The interaction of electron with the crystal lattice has the effect of — among other things — changing its effective mass. Here, of course, we can measure the mass of the electron with the crystal lattice interaction and without it, separately. Both are finite but different. The idea of renormalization in quantum field theory is similar and compares the hypothetical mass of the electron in the absence of all interactions with that of the real electron. The two key differences are the following: First, the bare mass can now never be measured since we cannot have an electron without interactions; second, in the bare mass has to be infinitesimally small for the observed mass to be finite because the corrections are infinitely large. The moral of this comparison is following: the physical reason for a parameter to change due to interaction is well founded; but the fact that it changes by an infinite factor makes one uncomfortable.

All this wizardry crucially depended on the fact that the infinite factors which arise in the calculation change *only* the parameters (like e.g. the mass, charge etc.) which were originally present in the theory. If this was not the case, the entire program would have failed and one should have given up constructing a relativistic theory of electromagnetic interaction at the quantum level. Clearly, the idea works only because the electromagnetic interaction has some special properties. Roughly speaking, this property is related to the fact that in quantum electrodynamics one is dealing with massive fermions (electrons) interacting via the exchange of massless vector boson (photon). By and large, such theories have the nice property that you can make sense of the infinities which arise in the perturbative calculations by renormalizing the parameters of the theory. They are called perturbatively renormalizable theories.

No one probably would have taken any of this too seriously except for the fact that it led to very concrete and observable effects. These effects could be tested in the laboratory and the agreement between theory and observation was so good that one had to concede that the shady procedure of dealing with infinities does have an element of truth.

4.4 Weak and strong interactions

In describing quantum electrodynamics as a prototype, we sort of took the easy way out in the sense that it is the simplest interaction to describe in relativistic quantum theory. But we do know that particles experience four different kinds of forces: Strong, Weak, Electromagnetic and Gravity. We need to worry about describing the other three interactions in a similar manner. Of these three, gravity is still unconquered. We simply do not know how to construct a relativistic quantum field theory of gravity and we will discuss this in Chapter 6. Here we will describe how the weak and strong interactions fit into the scheme of things.

There are several features which separate strong and weak interactions from the electromagnetic interactions, the most important one being the range of the interaction. Strong and weak interactions have a very short range and these forces fall rapidly at larger distances compared to a characteristic length scale. (In contrast, the electromagnetic force falls only as a inverse square power law at large distances.) It turns out that, in quantum field theory, the range of the interaction is inversely related to the mass of the quanta which mediates the interaction. If the quanta has a mass m, then the force of interaction will have a range which is about h/mc and the force will decrease rapidly at larger distances. If the mass of the exchange quanta is zero, the characteristic length scale becomes infinite and one gets a long range force. This is what happens in electromagnetism since the photon has zero mass. It is therefore obvious that to describe, say, weak interactions, we probably should use the exchange of quanta which are massive.

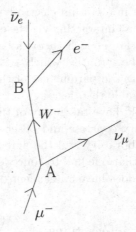

Fig. 4.10 The decay of the muon which occurs through the weak interaction and is mediated by a vector boson called W^-.

Figure 4.10, for example, could describe a typical weak interaction process involving the decay of a muon to an electron with the emission of two neutrinos: $\mu \to e + \nu_\mu + \bar{\nu}_e$. This is mediated by the exchange of a vector boson denoted by W^-. This vector boson has to be massive if weak interactions are to have a short range. What is more, it should carry a unit of negative electric charge (hence the minus sign in the superscript) in order to conserve charge at the vertex A. Since every particle should have an antiparticle, we expect a W^+ also to exist in the theory. One would have calculated the amplitude for the process in Fig. 4.10 by a procedure similar to the one used for Fig. 4.7.

There is, however, trouble ahead. Just as in the case of electrodynamics, here also one can draw other diagrams involving more complicated processes. Once again, when one sums up the amplitudes for all these processes, one encounters divergent expressions rendering the theory completely meaningless. To circumvent this, one has to use a procedure similar to that used in quantum electrodynamics and re-express all the observed parameters in the theory as renormalized expressions thereby taking care of infinities.

The crucial difference between electrodynamics and weak interaction is that in electrodynamics the vector boson which was mediating the interaction, viz. the photon, was massless; but in the case of weak interaction, we need the vector boson to be massive if the force has a short range. It turns out that the simplest kind of theories involving *massive* vector bosons are not renormalizable. That is, the relevant computations in such a theory will give you divergent, meaningless answers and one cannot make sense of them by changing the values of the parameters.

It turns out, however, that one can construct a more complicated theory which is indeed renormalizable and thus makes sense (usually called the Standard Model). But in this theory you need to have lot more ingredients to make sense.

To begin with, the theory will have 3 different kinds of vector bosons W^+, W^- and Z, all of which are required for mediating the weak interaction. The W's carry electric charge while the Z boson is neutral and, of course, they are involved in different kinds of processes. The existence of processes involving the neutral Z bosons was one of the key predictions of the theory, the verification of which bolstered one's faith in the model significantly.

The second ingredient in the theory is the existence of yet another kind of field, the quanta of which carries no spin. Such fields are called scalar fields or Higgs fields and the quanta of the field is called a Higgs boson. The existence of the Higgs boson is yet another key prediction of this theory — which has so far remained unverified in the lab. The scalar field is needed in order to provide masses to the particles in the theory. As we said before, if the vector bosons which are exchanged in the interaction are massless, then such a theory is a "good" theory. The trick in describing weak interactions is to start with a theory having massless particles but let them acquire mass through interactions by a process known as spontaneous symmetry breaking. This process is too complicated to be described here but you have already seen that interactions have a way of changing the masses of particles in quantum field theory and this is yet another example of the same. So you end up giving masses to the relevant particles thereby making the weak interaction one of short range without destroying the good features of the theory.

There is a remarkable bonus to this endeavour. When one works out all the maths and lets the dust settle down, the theory has not only three massive vector bosons, but also one massless, chargeless vector boson with exactly the properties of the photon. That is, you can start with a theoretical structure which describes at very high energies a combination of electromagnetic and weak interactions. As one moves down to lower energies, the theory — so to speak — separates into two: one involving 3 vector bosons which describe weak interactions and the other involving a photon describing the electromagnetic interaction. *We have managed to unify electromagnetic and weak interactions as a single unit!* Once again, the theory makes very concrete predictions which have been tested in the lab thereby confirming its veracity.

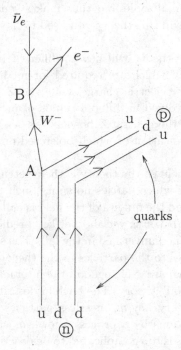

Fig. 4.11   The decay of the neutron in the beta decay. But since the neutron is made of quarks, the process, at the fundamental level involves the interaction of quarks as shown in the figure.

What is left is strong interaction and the story there is somewhat similar. The interaction is mediated by a set of vector bosons called gluons between particles built out of quarks. It is possible to build a theory for quarks and gluons in a somewhat similar manner. As an example of how fundamental physical processes operating at the level of quarks manifest in terms of more familiar particles, let us again consider the beta decay. The pictorial way of representing the beta decay is shown in Fig. 4.11. The process involves one down-quark getting converted into an up-quark through the emission of W^- particle at the event A. Further, at the event B the W^- particle is converted into an electron and an antineutrino. The other down and up quarks merely go for a ride in the process but the net effect is to convert a neutron into a proton emitting an electron and an antineutrino — which is precisely the beta decay. As a second example, consider the decay of the muon to the electron described earlier. The corresponding diagram is shown in Fig. 4.10. At each vertex (like A, B etc.) in these diagrams, the conservation laws for charge, flavour etc. do hold. For example, at A, one unit of negative charge arrives and leaves. You can also see that the antineutrino can be thought of as a neutrino propagating backwards in time along the general philosophy described earlier.

There is, however, a new problem which comes up at this stage. In the case of electroweak interaction, the particles which were interacting are normal particles seen in isolation in the lab. For example, the process described in Fig. 4.7 involves scattering of two electrons and we do see isolated electrons in nature. Similarly, the process in Fig. 4.10 involves muons, electrons and neutrinos all of which are seen in the lab. The strong interactions, however, involve quarks but we have never seen a single isolated quark in the lab. For some reason which is not completely understood, the quarks are seen only inside other particles like, for example, inside a proton. If one probes the interior of a proton in a very high energy scattering experiment, one does obtain (indirect) evidence for the existence of quarks inside the proton. What is more, these quarks which are inside the proton almost behave like free particles at high energies but all attempts to rip them out of the proton fail; one has to conclude that the quarks are confined inside other particles.

This confinement necessarily has to emerge as a result of the interaction between the quarks. Broadly speaking, this requires the strength of interaction between the quarks to decrease at very high energies but increase very rapidly at low energies — which is somewhat counterintuitive. The key reason for such a strange behaviour is again linked to the process of renormalization we have described earlier. We know that renormalization changes the value of the coupling constant (like, for example, the charge) which determines how strong the interaction is. Hence, it is very natural in quantum field theory for the strength of the interaction to depend on the energy scale of the interaction. In electromagnetism, for example, the coupling constant increases with the energy; so at low energies electrons behave as free particles and become strongly coupled at high energies. In the case of strong interactions the situation is exactly opposite. At low energies the coupling increases preventing the quarks from becoming free particles while at sufficiently high energies they behave like free particles inside, for example, a proton.

4.5   The secret life of the vacuum

We have seen that one of the key new features which arises when we bring together quantum theory and relativity is the possibility of creating and annihilating particles out of nothing. This "nothing" is what one would have normally called the vacuum state and it turns out that there is a lot happening in the vacuum state in quantum field theory. Since this idea will play a crucial role in our later discussion, we will take a closer look at the vacuum state and its physics.

To begin with, let us recall that what we call particles are just excitations of a field. To make this idea more concrete, we need a mathematical model describing the translation between fields and particles. In the simplest context, a field can be thought of as made of an infinite number of oscillators each oscillating with its own frequency. This is not a difficult idea to grasp because an electromagnetic wave is precisely that: At any given location in space the field oscillates in time with a given frequency. So the field can be thought of an assembly of oscillators each labelled by a particular frequency. (We are simplifying things somewhat here. Actually we also need to specify some information related to the direction in which the wave is travelling. But this description is adequate for our purpose.) When one proceeds from a classical wave to its quantum cousin, we have to introduce a quantum mechanical description for each of these oscillators. We have seen in Chapter 3 that the energies of such oscillators are quantized. An oscillator with frequency ω_1 will have energies $\hbar\omega_1(n_1+1/2)$ where n_1 is an integer including zero. If you have a bunch of oscillators with frequencies $\omega_1, \omega_2, \omega_3, ...$, its combined quantum state can be specified by giving a set of integers $n_1, n_2, n_3, ...$ which tells you the quantum state of each of the oscillators. But now the total energy of the system is given by the sum $(\hbar\omega_1 n_1 + \hbar\omega_2 n_2 + ...) + (1/2)\hbar(\omega_1 + \omega_2 + \cdots)$. This expression has a very nice interpretation. Let us ignore for the moment the energy of the ground state given by the second term $(1/2)\hbar(\omega_1+\omega_2+\cdots)$. (We will come back to it soon.) The remaining part of the energy $(\hbar\omega_1 n_1 + \hbar\omega_2 n_2 + ...)$ can be thought of as being provided by n_1 *particles* with energy $\hbar\omega_1$, n_2 *particles* with energy $\hbar\omega_2$ etc. That is, we can think of the excited state of a bunch of harmonic oscillators as equivalent to a system with particles having different energies. Each of the particle quanta should satisfy the relation $E = \hbar\omega$ which, of course, has been introduced very early in quantum theory. This mathematical connection between the vibrations of the field on the one hand and the equidistant energy levels of the quantized oscillator on the other hand is what ties together the field and particle concepts in quantum field theory.

This model also explains how particles can be created and annihilated in quantum field theory. The vacuum state of the quantum field will be the one in which there are no particles with any frequency and so we must have $n_1 = n_2 = \cdots = 0$. This would correspond to all the oscillators being kept in their respective ground states. To produce a particle of energy $\hbar\omega_1$ we just have to excite the corresponding oscillator from the $n_1 = 0$ to $n_1 = 1$ state. This will be an one particle state with, say, one photon or one electron, depending on the nature of the field we are dealing with. It is in this sense that we think of particles as excitations of the field.

What we want to explore now is not the nature of the particles but the nature of the vacuum state corresponding to $n_1 = n_2 = \cdots = 0$. We have already seen that this state has nonzero energy given by $(1/2)\hbar(\omega_1 + \omega_2 + \cdots)$. So, obviously, the vacuum state is not a zero energy state! As if that is not enough, a little thought shows that the energy of the vacuum state is actually infinite! This is because we need infinite number of oscillators to describe a field and when you add up an infinite number of $(1/2)\hbar\omega$ you get infinity. So quantum field theory tells you that the lowest energy state of the world has infinite energy.

This is pretty bad and what is much worse, nobody knows a fundamental solution to this problem. We will see in a later chapter that it leads to all kinds of trouble when one brings gravitational interactions into the picture. But if we ignore gravity, then there is a simple way out — which is to just ignore the infinite energy and pretend that it doesn't exist. You may not believe it but it works for an interesting and subtle reason. All physical interactions (except gravity which, as we mentioned, will lead us to trouble) really care only for energy differences and not for the absolute value of the energy. So you can subtract out $(1/2)\hbar\omega$ from every oscillator and pretend that the energy of the ground state is zero. As long as differences in energy are finite, everything will work out smoothly.

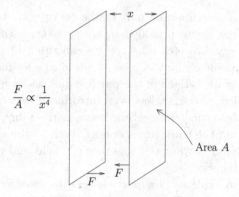

$$\frac{F}{A} \propto \frac{1}{x^4}$$

Fig. 4.12   Schematic configuration showing the Casimir effect. Two parallel conducting plates kept in the vacuum will attract each other with a force per unit area that varies as the inverse fourth power of the distance.

Does it mean that the ground state of a field has no observable effects? No, far from it. But in order to observe these effects, you have to compare two configurations each with infinite energy but with a finite difference between them. In such situations, the energy in the vacuum leads to observable effects. One such phenomena, called the Casimir effect, deserves special mention because of its remarkable simplicity. The effect can be described as follows. Consider two large plane metallic plates of area A and keep them parallel to each other with a separation x (see Fig. 4.12). It turns out that they will attract each other with a force per unit area which varies inversely as the fourth power of the separation. The simplest interpretation of this force is based on the energy density present in the vacuum (as illustrated in Fig. 4.13) which we will now describe.

As we said before, in the ground state of the electromagnetic field, all the oscillators are at $n_1 = n_2 = \cdots = 0$ configuration; but this does not mean that the field vanishes identically. The vanishing of the field will be equivalent to placing a particle at rest at the minimum of each of the potentials (which govern the field oscillations) thereby stating precisely its position and velocity. Since you cannot do that, the vacuum state of the field will have fluctuations in its field amplitude. Figure 4.13(a) is a schematic description of such a vacuum which can be thought of as having random fluctuations of the electromagnetic field corresponding to its ground state. All kinds of wavelengths and frequencies can exist for such fluctuations and the total energy of this configuration, as we have seen before, is infinite.

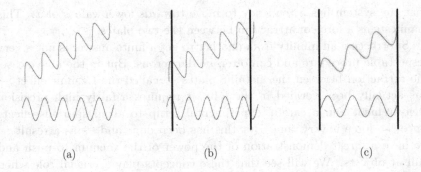

Fig. 4.13 Explanation for the Casimir effect in terms of the vacuum fluctuations. Part (a) of the figure shows some of the wave modes that can exist in the vacuum. When two plates are introduced into the vacuum, these modes, in general, will have nonzero electric fields at the conducting surfaces (see part (b)). Since this is not allowed, one has to exclude these modes and keep only the modes like the ones in part (c) in the presence of the plates. This shows that the vacuum is modified in the presence of plates. The energies of both vacuum states are infinite but their difference is finite leading to the Casimir effect.

Let us now introduce two metallic plates denoted by thick black lines in Fig. 4.13(b). Since the metallic plates are conductors, it is necessary that the electric field vanishes on them. (This is called a boundary condition in mathematical jargon.) This means that the configuration like the one shown in Fig. 4.13(b) cannot exist in the presence of the metallic plates. Only those kinds of fluctuations in which the field amplitudes vanish at the boundary of the conductor, like the one shown in Fig. 4.13(c), are now allowed.

This means that the introduction of the two metallic plates into the vacuum state of the electromagnetic field modifies the nature of the vacuum state. Of course, we have not created a photon — quanta of the electromagnetic field — in the process and hence the state is still a vacuum state devoid of particles. But the pattern of vacuum fluctuations have changed. The energy of the modified state is also infinite but now we have two vacuum states each with infinite energy. What is observable is the difference between these two energies and it turns out that this difference is finite. In fact, the difference is negative because less modes of vibration contribute to the second vacuum state as compared to the first. Calculations show that the energy difference per unit volume in the space between the two plates varies as $-(1/x^4)$ where x is the separation between the plates. This energy difference can be lowered by decreasing x which is the same as saying that the system has a tendency to move towards lower values of x. This manifests as a force of attraction between the two plates.

Subtracting an infinity from another to get a finite answer is not a very respectable procedure and can often be dangerous. But in the above case, the attraction between the metallic plates — called the Casimir effect — has actually been verified in the lab. It requires a fairly high precision measurement and a careful experimental setup to avoid spurious effects since the force is very tiny. But this has been done and — as a result — we have a direct demonstration of the power of the vacuum to push and pull at objects. We will see that these concepts play a crucial role when one attempts to incorporate the effects of gravity.

Chapter 5

Cosmos and the quantum

5.1 Structure of the universe

The discussion in the last few chapters would suggest that quantum themes apply in general to microscopic systems. It might therefore come as a surprise to you that one of the key application of quantum mechanics is in understanding the structure and evolution of the universe. The reason has to do with the fact that our universe is not static but is expanding (see Chapter 2). This implies that, in the past, it was hotter and denser and consequently microscopic physics played a crucial role in determining the physical processes in the early universe. What we see today is only the late time relic of events that took place early on in the evolution of the universe most of which are quantum mechanical in nature. We shall now turn to this fascinating story.

A view of the night sky might suggest that the universe is made of count-less billions of stars. But in reality, the stars cluster together as galaxies with a typical galaxy having about hundred to thousand billion stars. All the stars which you see in the night sky belong to our own galaxy called the Milky Way. There are billions of such galaxies strewn across the universe and the nearest large galaxy to Milky Way - called Andromeda - is about 2×10^{24} cm away. Astronomers use a unit called Megaparsec (Mpc) to talk about such large distances with 1 Mpc being equal to roughly 3×10^{24} cm. In this unit, Andromeda galaxy is about 0.7 Mpc. The farthest galaxies detected by our telescopes are about 5000 Mpc away from us and the mean distance between the galaxies is about 1 Mpc. Cosmologists are usually interested in structures at the scale of galaxies and bigger and you may say that galaxies are to a cosmologist what atoms are to a solid state physicist.

Interestingly enough, the galaxies are not distributed randomly in the sky but are clustered in a well defined manner. The Andromeda and Milky Way are the largest galaxies in a collection of about 30 galaxies called - rather unimaginatively - as the Local Group. A cluster of galaxies could contain anything from 30 to 1000 galaxies, all gravitationally bound together. For example, one of the largest clusters of galaxies called Coma Cluster, which is 300 million light years away, contains over 1000 galaxies. The distribution of galaxies in the sky shows a web like structure with large voids surrounded by sheets and filaments made of galaxies.

We said that the galaxies are to a cosmologist what atoms are to a solid state physicist. This analogy is quite useful in more than one way. When you look at the surface of a solid you usually find it to be pretty smooth with no sign of underlying granularity due to individual atoms. This is, of course, because one is perceiving the solid at length scales much bigger than the separation between two atoms. This idea translates directly to the study of our universe. At the largest scales, the universe appears to be smooth just as at scales much bigger than the interatomic spacing a solid appears to be smooth. Given the fact that the typical distance between the galaxies is 1 Mpc, we only need to consider regions of universe of size, say, 200 Mpc for the smoothness to manifest itself. A cubical box of size 200 Mpc kept anywhere in the universe will contain a statistically similar sample of galaxies.

Given this large scale smoothness, it is convenient to approach the subject of cosmology at two different levels. To begin with, one can consider the very large scale features of the universe by treating it to be filled with a fairly smooth fluid with specific properties. Having determined the large scale dynamics, one would like to come back to the formation and evolution of the individual structures like the galaxies. As you will see, this leads to a fascinating picture which is easy to grasp.

Let us begin by asking how the universe behaves as a dynamical system filled with matter having a fairly uniform density. Obviously this description is applicable only at scales, say, larger than 200 Mpc at which one can indeed ignore the granularity in the matter. Once again, in our analogy, this will correspond to studying the properties of a smooth solid with constant density. But unlike in the case of solids, th study of the smooth universe leads to a fascinating discovery. The universe at the largest scales is expanding! That is, any two galaxies in the universe separated by a sufficiently large distance will be moving away from each other with a speed which increases in proportion to the distance. Typically, two galaxies separated by, say, 500 Mpc will be flying apart with a speed of about 50000 km/s. This result, usually attributed to the cosmologist Hubble, can be deduced from direct observations of distant galaxies. The motion of the galaxy from us leads to a shift in the frequency of the light emitted by the galaxy. This effect, called the Doppler effect, is familiar to all of us - it is the same effect that changes the pitch of an oncoming train with respect to a stationary one. It is probably fair to consider the detection of the expansion of the universe as *the* discovery of 20th century. We saw in Chapter 2 that the description of the universe in Einstein's theory of gravity leads to precisely such a situation.

The fact that our universe is expanding implies that it was smaller, denser and hotter in the past and it evolves with time. In fact, calculations show that some time in the past - about 14 billion years or so ago - the volume of the universe was zero and its density infinite. This event was derisively called Big Bang by Fred Hoyle but the name stuck! This, in turn raises further questions: What happened before big bang? How does one understand scientifically the epoch of big bang?

We need to remember that, close to the "big bang", the universe was *infinitesimally* small - very much smaller than the size of an atom or even an electron! In trying to understand such a phenomenon we are extrapolating Einstein's theory of gravity, and the cosmological models based on it, to very *small* distances. However, the physics of the sub-atomic world needs to be described in the language of quantum mechanics incorporating the uncertainty principle. In the context of the universe, this will prevent us from determining simultaneously the size of the universe and its rate of expansion. It is, therefore, necessary to use a completely different description for understanding the physics of the universe close to big bang. Until we produce a quantum theoretical description of cosmology, one cannot understand the very early moments, close to the big bang. Unfortunately, we still do not have a good description of the quantum theory of gravity which incorporates the principles of Einstein's general theory of relativity and the principles of quantum mechanics. Until this is achieved — which is a frontier research area today — we cannot probe what happened to the universe near "big bang". (We will say more about this in Chapter 7.)

But we do understand a lot about the evolution of the universe since then. By tracing the expansion of the universe back in time one can try to understand the physical conditions that would have existed in the past. Such an exercise leads to a very fascinating picture of the early universe.

5.2   Thermal history of the universe

As we go into the past the universe will become more and more dense. When the universe was 10 times smaller, the matter density would have been 1000 times higher because the volume of the universe varies as the cube of its size. However, something curious happens to the energy of the *radiation* which exists in the universe. When the universe was 10 times smaller the radiation energy would have been 10,000 times higher (rather than just 1000 times higher). To understand the reason, it is best to think of radiation as made of a large number of photons. As the universe expands, the wavelength of each photon gets stretched. In the past, when the universe was 10 times smaller, the wavelength of each photon would have been 10 times smaller. Since the energy of a given photon is inversely proportional to its wavelength the energy of each photon would have been 10 times larger. The number density of photons also would have been higher by a factor 1000. (This factor, of course, is the same for matter density and photon number density.) Together the energy density contributed by the photons would have been 10,000 times higher when the universe was 10 times smaller.

This leads to an interesting conclusion. Since the energy contributed by radiation increases faster than the energy contributed by matter (as we go into the past), it is clear that at some sufficiently early phase, radiation energy will dominate over matter. To figure out when this will happen, it is necessary to know how much of the energy in the universe is in the form of radiation at present. It is found that most of the energy is distributed as a uniform background in the sky with the characteristics of a thermal spectrum. We saw in Chapter 1 that material bodies kept at some temperature will emit radiation with a characteristic spectrum (see Fig. 1.16) determined by the temperature alone. The background radiation in the universe has this particular shape with a characteristic temperature of about 2.73 K. It is as though the entire universe is enclosed by a box kept at this temperature. The thermal radiation at a temperature of 2.73 K will peak in the microwave region of the electromagnetic spectrum. Hence this radiation is usually called "microwave background radiation".

The energy density of the microwave radiation can be computed from its temperature. It turns out that this energy density (at present) is nearly 10000 times smaller than the energy density of matter. But since radiation energy density increases faster than matter energy density, it will catch up with matter density sometime in the past. Since radiation density today is down by a factor 10,000, it will catch up with matter when the universe was 10,000 times smaller. At still earlier epochs the universe will be dominated by radiation and not by matter.

Cosmologists use a convenient terminology based on the concept of redshift to speak about the early universe. We know that, as the universe expands, the wavelengths of photons are stretched, thereby introducing a redshift. When the universe expands by a factor 100, say, a wavelength λ would have become 100λ. The ratio between the increase in the wavelength $(100\lambda - \lambda = 99\lambda)$ to the original wavelength (λ) is called the "redshift z" (which is 99 in this case). This example shows that a redshift of z corresponds to an epoch when the universe was smaller by a factor $(1 + z)$. So instead of saying "when the universe was smaller by a factor 200 .... " we might as well say "when the redshift was 199 ....". For most practical purposes, when we are talking about the early universe, z will be sufficiently high for us to ignore the number 1 in $(1 + z)$. With this convention, we will often say that the radiation density was equal to matter density at a redshift of $z = 10,000$. This is just a way of talking about various epochs in the early universe but is extremely convenient.

As the density of the radiation increases, its temperature also goes up. When the universe was 1000 times smaller it also would have been 1000 times hotter, having a temperature of about $1000 \times 2.7 = 2700$ K. As we go further into the past, the size of the universe will become smaller and the temperature and density in the universe will increase further. According to theoretical calculations, the temperature and density of the universe would have reached infinite values at some finite time in the past, nearly 14 billion years ago. But, as we said earlier, one cannot have faith in a theoretical calculation which predicts infinite density and temperature and it is very likely that the theoretical models on which these calculations are based will break down well before such an epoch. Nevertheless, for the sake of definiteness it is convenient to call that moment — at which the density of a *theoretical* universe will become infinite — as the moment of "big bang" and measure the age of the universe from that moment. There are indications that the breakdown of Einstein's theory occurs at time scales smaller than 10^{-44} second when the temperature of the universe was higher than 10^{32} Kelvin. This is a *very very* high temperature and corresponds to an energy of 10^{19} GeV! At any reasonable temperature, like say 10^{10} Kelvin, or even 10^{20} Kelvin, Einstein's theory remains perfectly valid and can be used to make definite predictions. Using this theory, we can estimate the typical temperature of the universe at any given time. When the universe was about 1 second old, its temperature would have been about 10,000 million Kelvin. The average energy of a particle in the universe at this epoch is about 1 MeV, which is comparable to the energies involved in nuclear processes. The physical processes which operate at energy scales below about 100 GeV are well known from direct laboratory research. Hence we can certainly have faith in the predictions of the theory at these epochs. (We shall say something about the earlier epochs in Sec. 5.3.)

What was the material content of the universe when it was, say, 1 second old? Note that the temperature of the universe at this time was so high that neither atoms nor nuclei could have existed at this time. Matter must have existed in the form of elementary particles. The exact composition of the universe at that time can be determined by tracking back the contents from the present day. It turns out that the universe at this time was a hot soup of protons, neutrons, electrons, positrons, photons and neutrinos. For all practical purposes, we can ignore the existence of these neutrinos in our discussion. Positron (which is the antiparticle of electron; see Chapter 4) carries positive electric charge while electrons carry negative electric charge. This particle exists in the early universe because of a simple reason. The mass of the electron (or positron) is equivalent to an energy of about 0.5 MeV. When the universe is 1 second old, the temperature is about 1 MeV, which means that the energy of a typical photon is about 1 MeV. Since energy and mass are interconvertible, electron-positron pairs can be produced out of energetic photons at these high temperatures.

What about the number densities of each of these particles? This quantity can again be calculated from the present values of matter density and radiation density. To begin with, the total number of protons plus positrons should be equal to the total number of electrons. This is because the universe does not carry net electric charge. Each electron carries one unit of negative charge while each positron and each proton carries one unit of positive charge; only by keeping the total number of positrons plus protons equal to that of electrons will we be able to maintain overall charge neutrality. The number densities of protons and neutrons are approximately equal with neutrons being slightly less in number. As regards the photons, note that both the number density of photons and that of protons decrease in the same manner as the universe expands. So the *ratio* of these two number densities does not change when the universe expands. If we know this ratio today, it would have been the same in the past. From the energy density of microwave radiation and the density of matter, we can compute this ratio today. It turns out that the number of photons in the universe is much larger than the number of protons or neutrons; to every nucleon there are approximately 10^9 photons in the universe.

Having decided on the contents of the universe when it is 1 second old, we can try to determine the evolutionary history of our universe. As the universe expands, it cools; the average energy of the photons will decrease and soon the photon energy will drop below 0.5 MeV. When this happens, the photons will become incapable of producing the electron-positron pairs. The existing electron - positron pairs, however, annihilate each other producing radiation. When the universe cools to about 1000 million Kelvin, most of the positrons would have disappeared through their annihilation with electrons. We will now be left with a universe containing protons, neutrons, electrons and photons.

At this epoch, these particles exist as individual entities and not in the form of matter which we are now familiar with. To create the familiar form of matter, it is first necessary to bring protons and neutrons together and make the nuclei of different elements. Then we need to combine electrons with these nuclei and form neutral atoms. But we saw in Chapter 3 that the typical binding energy of atomic systems is in the range of tens of electron volts. At the temperature of few MeV such *neutral atoms* cannot exist. The binding energy of different atomic *nuclei*, however, is in the range of a few MeV. This suggests that it may be possible to bring neutrons and protons together and form the nuclei of different elements. However, this process — called "nucleosynthesis" — is not easy. The key difficulty, of course, is the fact that the universe is expanding very rapidly at this epoch. (When the universe is one second old, it will double its size every second; at present the universe is nearly 14 billion years old and hence it will double its size only in another 14 billion years. This shows why the expansion of the universe plays a major dynamical role in the early universe but not at present.) In order to bring particles together in one place we have to overcome the effects of expansion at least temporarily. This is quite impossible to achieve if we try to bring together several nucleons at once. For example, it is not easy to bring two protons and two neutrons together in one place to form the nucleus of helium, which is the second key element in the periodic table. Fortunately, there exists a nucleus called deuterium made of one proton and one neutron with a binding energy of about 2.2 MeV. As the temperature falls below this value, we can combine protons and neutrons and form deuterium nuclei. And then, by combining two deuterium nuclei together, one can form the nucleus of helium. This suggests that the first heavy elements to form out of the primordial soup are deuterium and helium.

It is important to ask how much of these elements are produced. Note that the binding energy of helium is 28.3 MeV while that of deuterium is only 2.2 MeV. So it is energetically favourable for all the deuterium to convert itself into helium. This is roughly what happens; but the expansion of the universe prevents the process from actually reaching the logical end. A tiny fraction (about one part in 10,000) of the deuterium survives being converted into helium. It turns out that more of deuterium survives, if the density of protons plus neutrons (called collectively as 'nucleons') in the universe is less. So by measuring the amount of deuterium produced in the early universe, we can constrain the amount of nucleonic matter present in the universe. Deuterium is a fairly fragile nucleus and is easily destroyed during stellar evolution. So any deuterium we see in the universe must have been a relic of the early universe, though not all of it would have survived. Thus observations give a lower bound to the amount of deuterium produced in the early universe (i.e., more could have been produced but not less). Observations suggest that this quantity is at least about 1 part in 10,000. Theoretical calculations, on the other hand, indicate that an abundance of 1 part in 10,000 is possible only if the total amount of nucleons in the universe contributes today a density of about 10^{-30} gm cm^{-3}. If the abundance of deuterium is more than one part in 10,000 then the amount of nucleons has to be still less. Thus you see that the *lower* limit on the observed deuterium abundance leads to an *upper* limit on the amount of nucleons present in the universe. As we shall see, this result has important implications.

Let us consider next the amount of helium which would have been produced in the early universe. Since each helium nuclei has two protons and two neutrons, one requires equal amounts of protons and neutrons to form helium nuclei. If the universe had precisely equal amount of protons and neutrons, then all of them can combine to form helium. If for any reason there was less number of neutrons, then these neutrons will combine with equal number of protons forming certain fraction of helium nuclei and the remaining protons will just stay as hydrogen nuclei. So the fraction of helium which is formed in the early universe depends on the ratio of the number densities of neutrons and protons.

Originally, in the very early universe, the number densities of neutrons and protons were equal. However, as time goes on, the number density of neutrons decreases in comparison with that of protons. This is because free neutrons in the universe undergo beta decay in which the neutron converts itself into a proton, electron and antineutrino. Because of this beta decay the number density of neutrons are continuously depleted vis-a-vis the number density of protons. By the time the universe has cooled enough to produce helium, there will be only about 12 neutrons to 88 protons in every collection of 100 nucleons (i.e., 12 percent of the nucleonic mass is in neutrons while 88 percent of the mass is in protons). These 12 neutrons will combine with 12 protons forming helium; the remaining 76 protons will end up as hydrogen nuclei. Thus about 24 percent of the mass will be in the form of helium nuclei with 76 percent of the mass will be in the form of hydrogen. We see the crucial role of quantum theory (via beta decay) in determining a result in cosmology.

In principle, one could try to produce heavier elements by further nuclear reactions. In practice, however, it is not easy to synthesize these atomic nuclei in large quantities, mainly because the expansion of the universe prevents sufficient number of protons and neutrons from getting together. In the periodic table, there are no stable elements with the atomic weight 8 (which could have been formed by combining two Helium nuclei) or with atomic weight 5 (which can be formed by combining a Helium nuclei with a proton). Detailed calculations show that only helium is produced in significant quantities. There will be very small traces of deuterium and other heavier elements.

The conclusion described above is a key prediction of the standard cosmological model. According to this model, only hydrogen and helium were produced in the early universe. All other heavier elements, which we are familiar with, must have been synthesized elsewhere. We now know that this synthesis took place in the cores of stars which have sufficiently high temperatures. (Stars manage to synthesize heavier elements because they could do it over millions of years at sufficiently high temperatures. In a rapidly cooling universe this is not possible.)

The role played by quantum physics in the study of the early universe is clear in all the above examples since none of the above features could be understood without using the principles of quantum theory. But once in a while cosmology also provides a new insight on particle physics as well. For example, one of the questions which puzzled particle physicists in the early days was whether there could exist more than 3 flavours of particles in nature. Can we have a fourth kind of neutrino, for example, in addition to the electron, muon and tau neutrino? Surprisingly, the study of primordial nucleosynthesis gives a clue on this issue and in fact constrains the total number of species of neutrinos to be three. This arises because, if nature had one more species of neutrinos, that would have also contributed to the energy density driving the expansion of the universe at the time of nucleosynthesis. This change in the rate of expansion of the universe changes the amount of helium that is synthesized. Observational constraints on the latter can be converted into a constraint on the number of species of neutrinos that could exist in the universe. (It is possible to arrive at the same result from particle accelerator experiments, for example, from the decay width of the Z particles. However, the first good constraint on the number of species of neutrinos came from cosmology.)

As time goes on, the universe continues to expand and cool. Curiously enough, nothing of great importance occurs until the universe is about several hundred thousand years old, by which time its temperature would have been about 3000 K. At this temperature, matter undergoes a transition from the so called plasma state (made of free charged particles like protons, electrons and helium nuclei) to ordinary gaseous state with the electrons and ions coming together to form normal atoms of hydrogen and helium. Once these atomic systems have formed, the photons stop interacting with matter and propagate freely through space. When the universe expands by another factor of thousand, the temperature of these photons would have dropped to about 3 Kelvin or so. *Hence we should see all around us today a radiation field with a temperature of around 3 Kelvin.* We mentioned earlier that the universe indeed contains such a thermal radiation at a temperature of 2.73 K. In 1965, two American scientists, Arno Penzias and Robert Wilson at Bell Labs stumbled on its discovery. As we shall see later, observations related to this radiation plays a vital role in discriminating between cosmological models.

5.3   Very early universe

The description of the universe given in the last section contains the best understood features of conventional cosmology. You would have noticed that we started the discussion with a universe which was one second old and carried it forward to several hundred thousand (about 400,000) years. Though this is an impressive feat, it leaves three questions unanswered: What happened before 1 second? What happened to the universe between 400,000 years and now? What will happen to the universe in the future?

Let us first discuss the more speculative of these questions, viz. what happened before about 1 second in the evolution of the universe? How did the universe get to this stage? The behaviour of the universe at earlier and earlier phases becomes more and more (theoretically) uncertain due to several reasons. As we go to very early phases of the universe the temperature is very high. This means that the elementary particles which constitute the matter in the universe are moving with very high energies. The interaction of particles and the laws governing their behaviour are uncertain when the energies involved are very high. Today, man-made accelerators in the laboratory can produce particles with energies of about 1000 GeV. When particles have energies significantly higher than 1000 GeV, we have no direct experimental evidence to understand their behaviour. Any theoretical model describing very high energy interactions is based on some broad symmetry principles rather than empirical facts. Given any such theoretical model it is possible to work out the consequences for the evolution of the early universe. In other words, how the universe behaved during its infancy depends crucially on the particle physics model adopted for high energy interactions. Until the latter is settled, the former cannot be predicted uniquely.

The usual procedure is to construct different possible particle physics models for high energies and work our their consequences for cosmology, with the hope that one will be able to constrain the particle physics models from cosmological results. Thus, in principle, the early universe provides a "poor man's particle accelerator" allowing one to probe high energy physics. In practice, however, this is not very easy because the predictions made by models may not be unique or may not be easily testable.

We pointed out in Chapter 4 that, at energies above 50 GeV or so, the electromagnetic and weak interactions merge together as a single interaction called "electroweak". It is hoped that at still higher energies — which could be as high as 10^{14} GeV — the strong interaction can also be "unified" with the electroweak force. We also saw in Chapter 4 that the forces acting between particles are mediated by photons for the electromagnetic interaction and W and Z bosons for the electroweak force (see Chapter 4; Fig. 4.6). At high energies, these particles will also be produced in large quantities and the dynamics of the early universe will be influenced by their existence. If the strong interactions are unified with the other forces at still higher energies, then it is necessary to have some more particles to mediate these unified interactions. These particles, (called X and Y bosons) are predicted to exist (at high energies) in some of these theoretical models. While the W and Z bosons are seen in the laboratory, we have no direct evidence for the X and Y particles. However, their properties can be predicted if the theoretical model is known.

There are some interesting new features which are brought into cosmology by such particle physics models. Consider, for example, the asymmetry which exists between matter and antimatter. Most of the universe is made of protons, neutrons and electrons but we do not see a corresponding number of antiprotons, antineutrons and positrons. This is surprising because at low energies, the interactions are symmetric between matter and antimatter. For example, the strong nuclear force which acts between a proton and another proton is the same as that which acts between a proton and an antiproton or between an antiproton and antiproton. That is, nuclear forces seen in the laboratory, do not care whether the particle is a proton or antiproton. If we start with an equal amount of matter and antimatter in some region of the universe, then these interactions cannot make them unequal since these interactions do not distinguish between matter and antimatter. So how is that our universe has matter but not antimatter? We can say "That is the way the universe is!" but that is no explanation at all! It would be nicer if we could start off the universe with an equal amount of matter and antimatter and induce an asymmetry through some dynamical process. Since the asymmetry already existed when the universe was, say, one second old, something which happened earlier on should have triggered this asymmetry.

It turns out that some of the models which unify the strong interactions with electro-weak interactions do require forces which break the symmetry between matter and antimatter. It may then be possible to start with a perfectly symmetric universe and arrive at a situation in which there is more matter than antimatter. Unfortunately, these models run into trouble when the details are worked out. Nevertheless, it must be remembered that the influx of particle physics ideas into cosmology has now at least provided us with a procedure by which such questions can be asked and possibly answered.

The models for high energy particle physics also predict different scenarios for the evolution of the earliest moments of the universe. Some of these scenarios contain pleasant surprises which help one to understand certain puzzling features of our universe. One such scenario which shot into prominence in the eighties is called the "inflationary universe". This scenario has helped cosmologists to look at the universe in an entirely new light.

5.4   Inflation

To understand the idea behind the inflationary universe and its importance, it is first necessary to recall some simple features of the expanding universe. If the universe is dominated by radiation, then it expands by a factor 2 when the time elapses by a factor 4, expands by a factor 3 when the time elapses by a factor 9 etc. Such an expansion is called a "power law" expansion and leads to certain conceptual difficulties.

The first difficulty is the following: In an expanding universe, physical processes cannot act as effectively as in a static universe. The following analogy will be helpful in understanding this feature: Suppose we divide a container into two parts and keep gases at different temperatures on the left and right halves. We now remove the partition and let the gases mix. Within a very short time the hot and cold gas will mix with each other and a common temperature will be reached inside the container. But suppose the walls of the container were expanding when the gases were trying to mix. Then the equality of temperature cannot be reached so easily. Regions in the centre of the container, where hot and cold gases are close to each other, will attain equal temperature after sometime. But whether the entire amount of gas will attain a constant temperature in an expanding container, depends on two factors: (i) the rate at which gas molecules can interact with each other and exchange energy; and (ii) the rate at which the container (and the gas) is expanding. If the rate of expansion is much higher than the rate of interaction, temperature equality will not be reached and the final state of the gas will be made of different patches with different local temperatures. It turns out that in the radiation dominated universe the expansion rate is such that no significant mixing can take place. In other words, if regions of the universe had different temperatures to start with, then they will never come to equilibrium and reach the same temperature. What is more, the patchiness of the temperature of the radiation must be visible in the sky, in the microwave radiation. Observations, however, show that the temperature of the microwave radiation is almost the same in all directions, suggesting that temperature in the early universe was the same at all spatial locations. In the conventional cosmological models this is a bit of a mystery.

There are two ways out of this difficulty. First one, of course, is to slow down the expansion tremendously — in the early phases of the universe — so that sufficient mixing can occur. This, however, is not possible because we know the rate at which the universe is expanding today and, from this value, we can calculate how fast it would have been expanding in the past. It is not slow enough for this idea to work. The second way out is the other extreme; that is, to let the universe expand very fast for a short time, after some amount of mixing has taken place. In this process, we take a small region of the universe which has some constant temperature and expand it by a very large factor to form the currently observed region of the universe. The universe *is* very patchy in such a model but the region which we observe today is made out of a single uniform patch, expanded by a large factor. Such models of the universe — which make the universe expand much more rapidly than the conventional models — are called the "inflationary models" and the rapid phase of expansion is called "inflation".

What drives the universe into such a phase of rapid expansion? Normally it is the energy density of the radiation that causes the early universe to expand. As the universe expands this energy density decreases and consequently the expansion rate also slows down. During the inflationary phase, however, the energy density remains constant (thereby maintaining a constant rate of expansion) even though the volume is increasing. In order to achieve this, the inflationary phase of the universe should contain matter with negative pressure. In normal systems which we are accustomed to (like a balloon filled with gas), pressure is a positive quantity. If the balloon expands, the energy content and pressure inside the balloon will decrease. But for a system with negative pressure the energy content will stay constant (or even increase) if the system expands! So, for inflationary models to work, we need some kind of matter with negative pressure. Normal matter, of course, does not have this property. But, quantum theory comes to one's rescue in this case. We saw in Chapter 4 that quantum theory establishes a correspondence between particles and wave fields. It turns out that certain models of particle physics use quantum fields which can have negative pressure (in the sense explained above). What inflationary models do is to arrange matters in such a way that these quantum fields dominate the expansion of the universe for a short span of time in the very early universe.

There is another significant fall-out from the inflationary models for the universe. The present day universe contains several inhomogeneous structures like galaxies, clusters etc. These structures can originate from the growth of small inhomogeneities present in the early universe. (We will discuss this in more detail in Sec. 5.6.) In order to develop a complete theory of structure formation, one has to also understand how these small inhomogeneities came into being in the first place. This problem is not adequately answered in the conventional cosmology. In fact the expansion rate of the radiation dominated universe is not conducive to producing inhomogeneities of the correct type. Inflation of the universe helps in this regard as well. The quantum fields which were driving the inflation will also contain small fluctuations in the energy density in a natural manner. These small fluctuations can grow and form the seeds for structures like galaxies. Given any particle physics model which can produce inflation, one can also compute the kind of fluctuations that would have been generated. This feature can help in understanding the formation of structures in the universe. Thus even the cosmic structure is of quantum origin!

How can we test such models? Rather, how can we be sure that the universe ever underwent an inflationary phase ? Since the inflation is supposed to have occurred at a very early phase of the universe (at about 10^{-35} seconds after the big bang!) it is not possible to test these models by direct observations. The typical energy of particles in the universe during inflation is far higher than the energies achieved in the laboratory; hence it is also not possible to verify the particle physics models leading to inflation directly in the laboratory. However, some indirect verification is possible. It turns out that generic inflationary models predict that the universe will have a density which is related to the expansion rate of the universe in a specific manner. This density is called critical density. (We will say more about this in the next section.) So if the observations conclusively show that the density of the universe is less than the critical density, then we can rule out the possibility that our universe underwent an inflationary phase. Each inflationary model also predicts a particular pattern of fluctuations for the inhomogeneities of the universe. The microwave background radiation contains information about these fluctuations. By comparing the observations of the microwave radiation with predictions of the inflationary models, one can validate or rule out these models. Recent observations (which we will discuss in Sec. 5.7) are completely consistent with inflationary models. Finally, most of the models of inflation also predict the existence of a uniform background of gravitational wave radiation in our universe. The intensity of these gravitational waves, however, is too low to be detected by the existing technology. But in principle, with an improved technology for gravitational wave detection, one should be able to test this prediction from inflation.

5.5   Dark matter

Let us next turn to the last question raised at the beginning of Sec. 5.3 viz., what will happen to the universe in the future? Will it go on expanding for ever? Or will it reach a maximum size and start contracting afterwards?

This situation is analogous to the behaviour of a stone thrown vertically up from the Earth. Usually, the stone will rise up to a maximum height and fall back on Earth. If we increase the initial speed of the stone, it will reach a higher altitude before falling back. This, however, is true only if the initial speed is below a critical speed called the "escape speed". If the stone is thrown with a speed larger than the escape speed, it will escape from the gravity of the Earth and will never fall back. In other words, if the gravity is strong enough, it can reverse the speed of the stone and bring it back; but if the gravity is not strong enough, the stone will keep moving farther and farther away from Earth. We can compare the expanding universe to a stone which is thrown from the Earth. If the gravitational force of the matter in the universe is large enough, the expansion can be reversed and the universe will eventually start contracting. Since the gravitational force increases with the amount of matter, the fate of the universe depends on the amount of matter present in the universe today. If the matter density is higher than a critical value, this gravitational attraction will eventually win over the present expansion and the universe will start contracting. The critical value of density which is needed for this recontraction to take place can be estimated if we know the speed with which the universe is expanding. It turns out that for our universe to contract back eventually, it should have a matter density of about $\rho_c = 5 \times 10^{-30}$ gm cm^{-3} or more. This density is called critical density of the universe. The ratio between the actual density ρ and critical density ρ_c is called the density parameter: $\Omega = \rho/\rho_c$. So we need $\Omega \geq 1$ to make the universe contract again.

So far so good. Now we need to know the actual mass density of the universe to determine the value of Ω. Is it higher than the critical density or lower? The future of the universe depends on the answer to this question.

A partial answer to this question is easy to obtain. We can make a fairly good estimate of the amount of matter which exists in the galaxies, clusters etc. as long as the matter is emitting electromagnetic radiation of some form: visible light, X-rays or even radio waves. Telescopic observations covering the entire span of the electromagnetic spectrum, along with some theoretical modelling, allow us to determine the amount of *visible* matter which is present in our universe. The density parameter contributed by this visible matter turns out to be about $\Omega \simeq 0.04$. So if all matter in the universe is visible, then the universe will expand forever.

Unfortunately, this is not the whole story. It turns out that all the matter which exists in the universe is *not* visible! In fact, it could very well be that up to forty percent of the matter in the universe is "dark" and does not emit any form of electromagnetic radiation. It is important to identify and estimate the amount of dark matter which is present in the universe since it is the dark matter which is governing the dynamics of our universe.

How do we determine the amount of dark matter present in the universe? Astronomers use different techniques for different objects but the basic principle is to look for gravitational effects which are *unaccounted* for by the visible matter. Consider, for example, the spiral galaxies. (These are galaxies in which most of the visible matter is contained in a plane in the form of a circular disc.) In the outskirts of such galaxies there exist tenuous hydrogen clouds orbiting the galaxy in the plane of the disc. If we see such a galaxy edge-on, then the hydrogen clouds at two sides of the galaxies will be moving in opposite directions - one towards us and one away from us. These clouds will be emitting radio waves with a wavelength of 21 cm which — because of the Doppler effect — will be redshifted at one end and will be blueshifted at the other. By measuring these shifts, we can determine the speed of these clouds. Knowing the speed and the location of the clouds, one can estimate the amount of gravitating matter present in the galaxy influencing the motion of the cloud.

Astronomers were led to a very surprising conclusion when they performed the above analysis. It turns out that galaxies like ours contain ten times more mass than as determined from the stars and gas. That is, there exists "dark matter" which is several times more massive than the visible matter in the Milky Way! Studies carried out for different kinds of galaxies have repeatedly confirmed this result: All galaxies have large dark matter halos around them with the visible part, made of stars etc., forming only a tiny fraction. Most of the gravitational force in the galaxy is due to the dark matter and hence the dynamics of the galaxy is mostly decided by the dark matter halo rather than by the visible matter.

What is this dark matter made of ? As we mentioned in Sec. 5.2, it is possible to put strict bounds on the number density of nucleons which can exist in our universe based on the primordial deuterium abundance. This bound implies that nucleons (which make up the ordinary matter) could at best only account for a few percent of the critical density. The amount of dark matter seen in various systems is far too large to be made of ordinary

matter. It is very likely that the dark matter is made of more exotic particles — particles that interact only very weakly and and have been produced
in copious quantities in the very early phases of the universe. Several particle physics models postulate the existence of more exotic particles called
'WIMPS' (Weakly Interacting Massive Particles) which could be far heavier
than protons, but much less abundant. There are currently several experiments in progress to detect such exotic particles, though none of them
have succeeded yet. A laboratory detection of a dark matter candidate
particle will revolutionize our understanding of the universe and establish
a remarkable connection between the physics at the smallest and largest
scales.

5.6 Structure formation

We saw earlier that our universe contains a hierarchy of structures from
planetary systems to clusters of galaxies. Between these two extremes we
have stars, galaxies and groups and clusters of galaxies. How did these
structures come into being?

This mechanism for the structure formation happens to be just our old
friend, gravitational force. The gravitational force can help the growth of
non-uniformity in any material medium. Consider the gravitational force
near two points A and B (see Fig. 5.1) with the actual density being slightly
higher in some places (like A) and slightly lower than average in locations
like B. Since the region around A has slightly more mass than the average,
it will exert more than the average force on the matter around. This will
make matter around A to contract thereby increasing the concentration
of matter around A. In other words, the density at regions like B and
C will decrease further while the density at A will go up. Now there is
even more matter around A; so the gravitational force near A will be still
larger causing still more matter to fall on to it etc. Clearly, this process
will make over dense regions grow more and more dense just because they
exert greater influence on the surroundings. This process goes under the
name "gravitational instability" and is similar to the social situation in
which the rich get richer and the poor get poorer. In describing the process
of gravitational instability, we have oversimplified matters a little bit. It
might appear that the lump at A grows essentially by accreting matter from
surroundings. Actually, there is another more important process; since
the region around A has more than average density, it will also contract

under its own weight. That is, the self-gravity of the lump at A will make it collapse further and further which will also increase the density of the lump. In fact, this process is really the dominant cause for the density to increase at A.

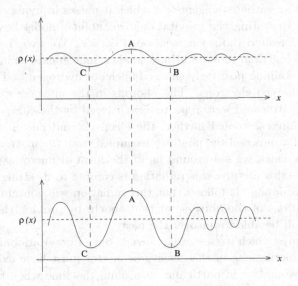

Fig. 5.1 The process of gravitational instability is schematically depicted in this figure. The top frame shows the variation of density of a gas in space. The dashed line represents the average density and the wiggles around the dotted line represent the actual density which is slightly higher in some places (like A) and slightly lower than average in locations like B and C. A high density region, like A, will contract under its own self-gravity and will also accrete matter from nearby regions like B and C. This process will make the lump at A denser and will decrease the density around it. So, at a later time, the density distribution will be as shown in the lower frame. This process, which leads to growth of density inhomogeneities, is called "gravitational instability".

The picture of gravitational instability depends on the existence of small wiggles in the density at some time in the past. If you start with a strictly uniform density — with no lumps like A — then, of course, it will remain strictly uniform in the future. So we need to have a mechanism which will produce wiggles in the density distribution in the early universe. As we saw earlier, the inflationary models can produce such small fluctuations. These wiggles could, of course, be quite tiny; gravity can make them grow eventually but gravity cannot operate if there were *no* wiggles to begin with.

This description attempts to build the entire universe starting from tiny density fluctuations which existed in the earlier phase. You could very well ask whether this model can be verified observationally or whether it is only a theoretical speculation. Fortunately, the gravitational instability model lends itself to a rather stringent test which it passes in flying colors!

The key to testing the gravitational instability model lies in the microwave background radiation mentioned earlier. We saw that when the universe was less than thousand times smaller, matter existed in the form of a plasma containing positively charged nuclei of hydrogen and helium and negatively charged electrons. The photons in the universe were strongly coupled to matter and were repeatedly scattered by the charged particles. When the universe cooled further, the electrons and nuclei combined to form neutral atoms and the photons decoupled from the matter. These are the photons which we see around us in the form of microwave radiation. According to this picture, this radiation is coming to us straight from the epoch of decoupling. It follows that this radiation will contain an imprint of the conditions of the universe at that epoch; by probing this radiation closely we will be able to uncover the past.

If structures which we see today formed out of gravitational instability, then small fluctuations in the density of matter must have existed in the early universe and — in particular — around the time when the universe was thousand times smaller. Since the radiation was strongly coupled to matter around this time, it would have inherited some of the fluctuations in the matter. What does one mean by fluctuations in radiation? Since the thermal radiation is described essentially by one parameter, namely, the temperature, fluctuations in the density would correspond to fluctuations in this temperature. Calculations showed that this temperature difference is extremely tiny — about one part in a hundred thousand! But it is vital to look for these temperature fluctuations to verify the theoretical models.

5.7   Ripples in radiation

In the late seventies and eighties, several experiments attempted to measure these temperature variations. The water vapour in the atmosphere absorbs and emits radiation in the microwave region of the spectrum (in which one is trying to make measurements) and this interference makes it very difficult to carry out these experiments from the ground. To have a fighting chance, one has to carry out experiments at exceptionally high or dry locations.

Even then, these experiments could only reach a sensitivity of few parts in a hundred thousand and did not see any deviation. The microwave radiation looked quite uniform till about 1992.

The first breakthrough came when NASA put into orbit a satellite, called the Cosmic Background Explorer (COBE), specifically designed to look for temperature deviations in the microwave radiation. Analysis of data collected by COBE in 1992 showed that the radiation field does have temperature variations of expected magnitude: about one part in a hundred thousand! COBE measured the CMBR over the entire sky with an angular resolution of about 7 degrees. Temperature anisotropies at these scales were expected to come from the gravitational potentials of the largest scale structures in the Universe. Roughly speaking, the photons which come from deeper gravitational potential wells will lose more energy compared to those coming from shallower potentials, which will lead to a temperature anisotropy that is roughly the same at all large angular scales. (This is called Sachs-Wolfe effect after the two scientists who first worked it out in 1967.) Over the past 16 years several other experiments have measured these temperature variations and reconfirmed the COBE observations – the most recent being Wilkinson Microwave Anisotropy Probe (WMAP).

What are the implications of this discovery for cosmology? To begin with, it shows that the theoretical models about the formation of the structures in the universe are basically sound. The universe was indeed smoother in the past but not completely so. The small departures from uniformity are what have grown into galaxies we see today.

Second, the more recent experiments has thrown some light on another key issue in astrophysics: the nature of dark matter. Obviously, one would very much like to know whether the dark matter is made of same building blocks as visible matter (made of neutrons and protons which contribute most of the mass of atoms; usually called baryonic matter) or whether it is made of any other kind of elementary particle. Again COBE results have something to say about this: The amount of temperature variations detected in the microwave radiation lends support to the idea that dark matter is non-baryonic. Calculations suggest that the temperature fluctuations would be nearly a hundred times larger than what was detected, if the dark matter was made of the same material as ordinary matter. So the dark matter is made of particles which are different from normal matter.

The story is actually more fascinating than that. Even before the COBE results came, cosmologists believed – based on the observations of spiral galaxies etc. mentioned in the last section – that most of the matter in the universe was dark and was made of some weakly interacting, massive, elementary particles. The most popular model at the time of COBE was something called cold dark matter (CDM) model. This model is based on a universe with critical density with most of the density contributed by dark matter particles. COBE results added an intriguing twist to this tale when combined with other observations. The theoretical analysis of COBE data suggested that the dark side of the universe was actually made of two components. One was the the standard dark matter but the other one was distributed very smoothly over the large scales; that is, even the dark sector of the universe should contain at least two different components. COBE could not throw more light on the dark side but subsequent cosmological observations (like WMAP) have now led to an intriguing composition for the universe: two-thirds of the dark sector is made of an exotic component with negative pressure — now called 'dark energy' and one third is made of cold dark matter. We shall now discuss this new twist to the tale.

5.8 Dark energy

One key result that emerged from the recent observations of cosmic microwave background radiation (by e.g. WMAP) is the following. The total energy density of our universe is precisely equal to the critical density but nearly seventy per cent of it must be made of a very smooth dark component which exerts negative pressure. It is this component which has been christened dark energy and for the last several years cosmologists have been trying to come to grips with it.

The two strange features which distinguish the dark energy from the dark matter are the following: First, dark matter behaves like any other material particle, is affected by gravity and clusters gravitationally. This is the reason why we have halos of dark matter around galaxies and clusters of galaxies exerting gravitational influence on the galaxies. In contrast, the dark energy seems to be distributed in a completely smooth manner with almost no wiggles in the density distribution. Second, and more strange, is the fact that dark energy exerts negative pressure on its surroundings. This is exactly the situation we had during the inflationary phase. But the amount of energy and the timescales involved are quite different. Recall that if a balloon was filled with a gas exerting negative pressure, then the expansion need not necessarily decrease its energy content. As the region expands, the energy content can remain constant or even increase if the pressure is sufficiently negative. This is indeed esoteric but believe it or not, observations show that seventy per cent of the matter in the universe has such an esoteric behaviour.

How could one test this idea directly? Fortunately, there is a way. A universe filled with normal matter will decelerate as it expands. If you imagine the expanding universe with a moving car, you can think of three different scenarios for its motion. The car can move along in a particular direction with a steady speed, it can move with an ever increasing speed (acceleration) or it can move forward with decreasing speed (deceleration). Normal matter in the universe - which has no negative pressure - leads to an expansion with deceleration. On the other hand, the characteristic of matter with negative pressure is that it will lead to an accelerating universe in which the speed of the expansion will keep increasing with time. (Again, this is what happens in the inflationary phase causing the universe to expand very rapidly.) So all we need to determine is whether the expansion rate of the universe has increased over time in order to determine whether the universe is dominated by dark energy with negative pressure. Fortunately, this is possible because cosmologists are a blessed lot who can see into the past. Since light takes finite time to go from one point to another, you never observe any object as it is at that instant. For example, when you look at the Sun, you see it as it was about 8 minutes ago since it takes that much of time for the light to come to you from the Sun. Similarly, if you observe a galaxy which is 10 billion light years away, you are seeing it as it was 10 billion years ago! So by observing a sequence of objects at ever increasing distances, one can figure out how the universe was behaving at earlier and earlier times. In particular, one can settle purely observationally whether the universe is accelerating or not.

This was attempted using certain powerful beacons of light called supernova, which are produced when, at the end stages of their life cycle, some stars die explosively and brighten up enormously over a period of time. Such supernova, because of their brightness, can be detected up to a very large distance and can be used to determine whether the universe was accelerating or not. These results, obtained over last several years, suggest that the universe is indeed accelerating. Figure 5.2 shows the observational situation in a dramatic manner. The vertical axis of the figure gives the rate of expansion of the universe and the horizontal axis indicates the size of the universe. The left part of the curve shows that, as the universe expands, the rate of expansion was initially decreasing. This is the phase of deceleration which is expected if the expansion of the universe is dom-

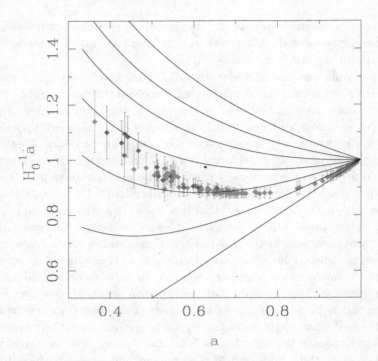

Fig. 5.2 The data from supernova shows that the universe is accelerating in the recent past. The figure shows the speed of expansion of the universe as a function of its size. One sees that the speed was decreasing initially (viz., the universe was decelerating) reaches a minimum an then starts increasing (viz., the universe was accelerating) in the recent past.

inated by normal matter with positive pressure; as the universe expands, the energy density of the matter decreases reducing its efficacy in driving the expansion. As a result, the universe expands at a lower and lower rate. The situation changes sometime in the past when the universe was about (3/4)th of its present size. The figure shows that around that time the rate of expansion reached a minimum and subsequently it started increasing. That is, the expansion of the universe started accelerating. This, in turn, is expected if the universe is dominated by matter with negative pressure. The energy density of such exotic matter can remain constant even as the universe expands, driving it to faster and faster rate of expansion. In fact the observations show that the universe indeed has about seventy per cent of the matter exerting negative pressure exactly as suggested by the microwave background experiment. The fact that two completely different observations have led to the same model of the cosmos has given enormous credibility to these ideas which otherwise would have been treated with much more skepticism.

The key properties of the dark energy fluid can be described in terms of what is usually called the equation of state parameter. This parameter, denoted by w, describes the ratio between the pressure and energy density of the dark energy. Observations suggest that w is a negative number and is very close to -1. It has to be negative so that the pressure will be negative for a positive value of energy density.

So, what exactly is this dark energy? The frank answer will be that: "We do not have the faintest idea!" But there are several exciting and imaginative possibilities that have been suggested by the researchers and this is indeed a key research area in cosmology during the last one decade or so. One way of modelling the dark energy is to think of it as an exotic fluid with negative pressure. It turns out that you can model such a fluid mathematically by using a scalar field. As we said before, a scalar field is just like an electromagnetic field with two crucial differences. First, it is represented by a scalar quantity which is a mathematical way of saying that it is a single number at each event in spacetime, unlike, for example the electric field which is a vector and requires three components (that is three numbers) to be specified at each event. The second — and probably rather disappointing — feature of the scalar field is that no one has seen a scalar field in the real world in contrast to, say, an electromagnetic field. We, therefore, have to postulate its existence just to explain the existence of dark energy seen in the cosmological observations.

Depending on the properties of the scalar field, one can arrange for the dark energy to behave differently at different epochs in the evolution of the universe. Such models are somewhat esoterically called quintessence models. The current observations do not give any evidence for variation in the properties of the dark energy fluid as the universe evolves. For example, if we keep the equation of state parameter w at a fixed value of -1, say, then one can still explain all the observational data we have which describes the universe from the epoch at which it was about six times smaller than the current size. At the same time, the observations are not yet tight enough to completely rule out the variation in the value of w to some extent. The quintessence models generically will have varying w. One of the great challenges facing the future observations is to determine whether w changes with time or not. At present, the situation is completely open.

You may be wondering why we chose the value $w = -1$ instead of some other negative number, say, $w = -0.75$. It turns out that $w = -1$ is indeed special and in fact it can be realized without postulating any quintessence or scalar field. It arises quite naturally if you add a term to Einstein's equations called the cosmological constant which we mentioned in Sec. 2.6. This particular term — which just modifies Einstein's equations — behaves exactly like a fluid with a pressure p and energy density ρ related by an equation $p = -\rho$, which, of course, is same as $w = p/\rho = -1$. This simple choice for dark energy is something suggested by Einstein himself years back in a completely different context. As we mentioned in Chapter 2, when Einstein found that the equations describing the cosmological model in his theory did not allow a static universe which lasted from ever lasting to ever lasting, he tinkered with his equations by adding an extra term (called the cosmological constant) in order to obtain a static universe. The term he added did not quite result in a stable static universe and very soon others pointed out that even with the cosmological constant, the universe will continue to expand. If the term which Einstein had is present, it will lead to the kind of accelerated universe that we do see.

The cosmological constant would have been an ideal solution to what we see in the universe except for one serious problem. The difficulty is in understanding the actual numerical value of this cosmological constant. It turns out that this constant has to be very small and should have a very finely tuned value in order to describe the universe which we see. A small change in its value will lead to universes which are totally different from the one we are living in. It has been a challenge to understand why the cosmological constant has the value it has. This perceived difficulty has led researchers to try out other possible candidates for dark energy, each more bizarre than the previous, but none of them has proved to be more successful than the cosmological constant.

There is another intriguing side to the cosmological constant which arises from quantum field theory. We saw in the last chapter that the vacuum state of quantum field theory is not a "mere" nothing but is bristling with fluctuations of all kinds of fields. It is conceivable that these fluctuations carry some energy density and can exert pressure. Recall that pressure is nothing but the force per unit area exerted on a surface which is precisely what we found the vacuum is capable of exerting when we discussed the Casimir effect in Sec. 4.5. This raises the question: What is the relationship between the pressure and energy density of the vacuum?

Incredibly enough, it turns out that the equation of state for the vacuum is $p = -\rho$ which is identical to the one mimicked by the cosmological constant. Proving this fact is mathematically intricate but one can provide a qualitative understanding along the following lines. Consider a rectangular box closed on one side by a movable piston with all other walls remaining rigid. Let the region enclosed by the walls of the box and the piston be in vacuum state. Suppose the pressure p_{vac} exerted by the vacuum makes the piston move slightly thereby changing the enclosed volume by V. It turns out that this will change the energy of whatever is contained inside (in this case just the vacuum, treated as some kind of a physical medium) by the amount $-p_{vac}V$. This requires us to attribute an energy $E_{vac} = -p_{vac}V$ to the vacuum state itself so that the energy density $\rho_{vac} = E_{vac}/V = -p_{vac}$. This suggests that one can probably think of the cosmological constant as arising due to the vacuum fluctuations itself. Of course, an equation of state for the vacuum of the form $p_{vac} = -\rho_{vac}$ has the elementary solution $p_{vac} = -\rho_{vac} = 0$ which is equivalent to saying that the vacuum has no energy density or pressure. But if we postulate that vacuum has a nonzero energy density, then its equation of state matches with what the dark energy observations suggest.

To make this idea work, we need to compute the energy density of the vacuum from first principles and see whether it matches the observations. Unfortunately, no one has been able to do this in a reliable manner. The difficulty is partially what we alluded to in the last chapter. The computation of the energy associated with the vacuum leads to a divergent result; that is, an infinitely large number. In the case of the Casimir effect which we discussed in Sec. 4.5 we got away with this by computing the *difference* between energies of the vacuum with and without the capacitor plates. But now we want to compute the absolute value of the vacuum energy — which is far from easy. Further, we now need to compute the gravitational effects of the vacuum energy since this is what the cosmological observations essentially measure. Gravity will not only be influenced by the absolute value of the vacuum energy density (assuming it is nonzero) but will also modify it; so we need to compute everything self-consistently in a curved spacetime. This leads to new difficulties some of which we will allude to in the next chapter.

Chapter 6

Matters of gravity

Earlier on, in Chapter 2, we described several features of gravity and noted that it is somewhat of an odd-man-out amongst the forces of nature. In Chapters 3 and 4 we introduced the principles of quantum theory and showed how they are universal in the sense that they are applicable in the description of all the laws of nature. Just as laws of electrodynamics got modified to a description based on quantum electrodynamics — when the principles of quantum theory are brought in — one would have expected the description of gravity to get modified into something, which we could call quantum gravity, when we bring in the same principles.

Nowhere do the peculiarities of gravity play a greater role than in frustrating the attempts of the physicists to combine the principles of quantum theory and gravity. As a result, we do not yet have a quantum theory of gravity but only a plethora of candidate models and imaginative ideas. There are, however, strong indications that bringing together the principles of quantum theory and gravity will usher a new revolution in physics possibly more dramatic than the relativistic or quantum revolutions and thus, will enrich our knowledge of nature. In this chapter, we shall describe several aspects of these issues.

6.1 This peculiar thing called gravity

In the good old days of Newton, gravity was just a force like the electromagnetic force. Forces act between particles and make them move in space and time. The special theory of relativity changed the way we look at space and time but did not really modify our fundamental ideas of dynamics. The description of electromagnetic force, for example, is perfectly consistent with the principles of special relativity. The force acted between charged particles and made them move. How exactly the particle moved was described by an equation which was different in Newtonian mechanics compared to special relativistic mechanics; but once that difference was taken care of, it was business as usual and one could go ahead and calculate things.

As we saw in Chapter 2, this idea did not work with gravity. One could not combine the principles of special relativity with the gravity and still describe the latter as a force which was acting between particles. Instead, the correct description of gravity had to be based on the concept of a curved spacetime. Material bodies — or, more generally, any form of energy — curves the spacetime around it. Other bodies move in a straight line in this curved geometry rather than "feel the gravitational force". As we described in Chapter 2, a straight line in a curved space (like the surface of a sphere) will appear to be a curved line if we use standard 'straight' coordinates adapted to flat surfaces. In this picture, the Earth going around the Sun is actually moving in a straight line in the spacetime curved by the Sun! It is obvious that the ideas which we use to provide a quantum description of a *force*, like electromagnetic force or nuclear force, will not work in such a bizarre context involving a spacetime which itself is curved. This is precisely what happens.

There are also several other ways in which gravity is different from other forces all of which have implications for any attempt to produce a quantum theory of gravity. If we naively think of the four different forces described in Chapter 4 as electromagnetic, gravitational, weak and strong (ignoring for the moment that electromagnetic and weak forces have already been unified) then, gravity is a long range force like the electromagnetic force. This is in contrast with weak and strong forces which have short ranges. We saw in Chapter 4 that, in the language of quantum theory, the range of the force is related to the mass of a particle which is exchanged. The electromagnetic force arises due to the exchange of photons between charged particles and similarly, one would have thought the gravitational force should arise due to the exchange of a particle, usually called a graviton, between any two material bodies. To have long range, both the photon and the graviton should be massless.

But the electromagnetic force acts only between charged particles and hence its long range is not of much practical value. In other words, bulk matter in the universe is mostly neutral and has no electromagnetic force acting between them. For example, the electromagnetic force exerted by Earth on Moon is utterly negligible compared to the gravitational force even though the electromagnetic force of attraction between a proton and electron is much larger than their gravitational force. This is because Earth and Moon carry negligible (net) electric charge but a large amount of mass which leads to this situation. As a result, the gravitational force becomes more and more dominant as we move to larger and larger scales in the cosmos. At the largest scales, like those of galaxies, gravity is essentially the only force we need to reckon with. At these scales, we had already seen that gravity is not a force but is best described as due to curvature of spacetime.

The last fact produces a serious conceptual difficulty when one tries to understand gravity using ideas similar to those in the successful paradigm of, say, quantum electrodynamics. Much of the success of quantum electrodynamics arises due to the fact that one could deal with the effect of electromagnetic interaction by using a series of approximations. The mathematical idea underlying this can be translated roughly as follows: In quantum electrodynamics, one could start with a notion of, say, two electrons which are blissfully unaware of each other. We then add the effect of what will happen if they exchange one photon between them; then we add the effect of, say, two photons being exchanged between them or the effect of the photon which is being exchanged behaving like a electron-positron pair for a brief period of time. Each of these effects can be thought of as a small correction to the original situation and can be computed in a controlled manner. Roughly speaking, this is like adding to the number 1, a series of "corrections" like 0.1, 0.01, 0.001 etc., each of which is small compared to the previous one. If you stop adding these numbers at some stage, you still get an answer which is reasonably accurate and the more numbers you add the better your result is. But remember that this whole idea worked only because the initial state of the system could be taken as the one with no interaction between the two electrons.

It is now easy to see why this approach is conceptually problematic for gravity. If you think of gravity as curvature of spacetime, then the absence of gravity will be represented by a flat spacetime. Taking a cue from quantum electrodynamics, we then need to add effects of material bodies exchanging gravitons in order to compute quantum gravitational effects. If you want to study the effect of one electron scattering off another electron due to their gravitational interaction, this is a good starting point. (It turns out that the idea doesn't work even in this case — for very important reasons — but we will discuss that a little later.) But we are almost never interested in such questions in the study of gravity. As we said before, gravity dominates the dynamics at very large scales and at these scales it is simply incorrect to start with the assumption that spacetime is flat and then try to add the effects of gravity little by little. In technical terminology, one says that the interesting questions in gravity are non-perturbative in nature. So we really need a different paradigm just because gravity in the bulk already produces a very strong effect at the classical level.

But suppose for a moment we are willing to ignore this complication and want to push the idea of the gravitational force between two electrons as arising due to the exchange of gravitons. Can we compute this effect like the way we computed the corresponding effect in quantum electrodynamics? The answer is a big NO! In fact, when we try to do that, we get utter nonsense. This fact is so important in understanding the problems of quantum gravity that it is worth looking at it more closely. The problem which arises here is very similar to the one which we mentioned in the case of quantum electrodynamics. It just turns out that in the case of QED (as well as in the electro-weak theories), there is a mathematical trick by which this problem could be pushed under the rug. The trick does not work for gravity. Let us see why.

The lowest order process in which two electrons interact through an exchange of a graviton will look schematically like the one shown in Fig. 6.1(a). This is very similar to Fig. 4.7 of Chapter 4 which described the electromagnetic interaction between two electrons arising due to the exchange of a photon. But, at the next level, one has to add processes like the one described in Fig. 6.1(b),(c). In Fig. 6.1(b) the graviton becomes a pair of virtual particles (represented by the circular loop in the centre) while being exchanged between the two electrons. A similar effect arose even in quantum electrodynamics (see Fig. 4.9) in which a photon can transform itself into an electron-positron pair. The process described by Fig. 6.1(c) has no analogue in quantum electrodynamics but occurs in gravity. This process shows that a graviton can transform itself into virtual gravitons as well. For this to happen, a graviton has to interact with itself. We know that the source of gravitational field is essentially the energy and since gravity itself carries energy (at least in the weak field limit at which we are describing the phenomena), it is no surprise that gravity couples to itself. In quantum language, the graviton couples to itself resulting in processes like in Fig. 6.1(c). In contrast, the electromagnetic field has as its source the electric charge but the photon does not carry any charge. So the electromagnetic field does not couple to itself unlike the gravitational field. This is why there is no analogue of Fig. 6.1(c) in the case of quantum electrodynamics. Mathematically, one says that electrodynamics is a linear theory while gravity is non-linear.

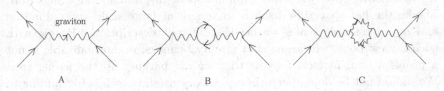

Fig. 6.1   The diagrams representing some processes which contribute to the gravitational scattering of two particles. (a) The simplest process in which the particles exchange one graviton. (b) The graviton transforms to a virtual particle-antiparticle pair in the intermediate stage (c) The graviton transforms into a pair of virtual gravitons in the intermediate stage; this is possible because the graviton couples to itself.

We can try to calculate the amplitude for physical processes represented by diagrams like Fig. 6.1(b),(c). When we do that, we again get infinite answers just as in the case of quantum electrodynamics. But in the latter we saw that there is a systematic procedure for extracting finite results out of potentially infinite contributions. This procedure, or rather the mathematical trick, fails in the case of gravity. You will recall that this trick relied on changing the values of various parameters in the theory — like the mass, charge etc. In the case of quantum electrodynamics, we start with the theory having a certain number of parameters and by adjusting their values we can obtain finite results for all the physically relevant parameters. In the case of gravity, we find that we cannot do this. If we start with a certain number of parameters, say two, and compute a process like the one shown in Fig. 6.1(c), we are forced to the conclusion that another new, third, parameter needs to be introduced in the theory. When we redo everything with the new theory having 3 parameters we find that it generates extra diagrams with infinite contributions which can be accommodated only if we had a fourth parameter and so on. In such a theory you cannot predict anything because you need to know the values of an infinite number of parameters to describe the physical processes. In technical terminology, such a theory is called non-renormalizable. Gravity is a prime example of non-renormalizable theory while the theories describing electromagnetic, weak and nuclear forces were all renormalizable.

There is an important caveat to the above comment. Recall that the above approach was based on starting from the flat spacetime and adding the effects of gravity order by order like the process of adding 0.1, 0.01, .. etc. to unity. As we mentioned earlier, we are not quite sure this is the conceptually correct procedure in the case of gravity. Maybe one shouldn't try to describe gravity starting from flat spacetime and adding things to it. Maybe the flat spacetime itself is an emergent large scale approximation and at microscopic scales we need a different description. People with such a view point will argue that gravity being non-renormalizable is not a problem and, in fact, tells us that we are barking up the wrong tree. We should not be doing perturbation theory at all (which is like adding to unity small numbers one at a time) but instead look at gravity completely non-perturbatively.

Just to show you what is involved, let us consider a more familiar situation in which such a feature arises. A solid body like an iron rod, for example, looks completely smooth and continuous in our everyday experience. You can change the shape of such a body by bending it or twisting it; you can make it vibrate by tapping at one end with some force etc. All these dynamical phenomena can be understood using what is called the theory of elasticity. Engineers use this all the time if they are going to use iron frames to build a bridge or use a beam to support a large building. They will need to know how the iron rod will behave when you apply some load to it and this can be understood in terms of the elastic properties of the material. Essentially these are just a set of a few numbers and for most elementary applications, one just needs 3 numbers. All this goes to show that you can develop the bulk properties of an iron rod by assuming that it is a continuous, smooth object with some density and certain elastic properties.

But we know that — at a fundamental level — this is simply not true! All matter, all solids including the iron rod in question, are made of atoms arranged in some specific manner at microscopic level. At the scale of the atoms, the iron rod is far from smooth or a continuum. It is made of discrete structures with lots of empty space in between and when you take into account the fact that even inside an atom most space is empty, you realize that the description of a solid with a smooth density is an approximation valid only at large scales.

The laws which govern the behaviour of the iron rod at the microscopic scale are quite different from the laws of elasticity which the engineers use. The microscopic phenomena are governed by the laws of quantum physics and ideas like density of the solid or the elastic constants are quite inappropriate at that level. But when one considers a large collection of smoothly arranged atoms, the phenomena of elasticity arises with a life of its own.

It is quite possible that gravity is somewhat like elasticity. The description of gravity in terms of a curved spacetime, for example, could be an approximate description valid at large scales just as the description of a iron rod in terms of the theory of elasticity is valid at large scales. In the case of a solid, large scale simply means that you are interested in sizes much larger than the mean separation between the atoms in the iron lattice. The typical separation between such atoms is about one hundred millionth of a centimetre. So, if you are looking at a cubical block of iron one centimetre on its side, it will contain typically about 10^{24} atoms. At these everyday dimensions, one can ignore atomic physics and use the theory of elasticity to describe the block of iron.

What does 'large scale' mean in the context of gravity? Fortunately, nature has given a rather unique answer. From the three fundamental constants G (Newton's gravitational consant), c (speed of light) and $\hbar$ (Planck's constant), one can obtain a lengthscale $L_p \equiv (G\hbar/c^3)^{1/2}$ called the Planck length. Numerically, it is about 10^{-33} cm which is *very very* small. To understand just how small it is, you only need to realize that the size of the visible universe is only about 10^{28} cm. So, if you start with an everyday dimension of 1 cm and multiply it by a factor ten 28 times, you reach the size of the visible universe. In the other direction, if you start with 1 cm and divide it by a factor ten 28 times, you do reach a very tiny size but this size is still much larger than the Planck length! You have to divide by ten 33 times to reach Planck length. Thus, in terms of powers of ten, Planck length is farther away from everyday scale than the edge of the universe is! We all know the kind of esoteric phenomena which arise in the description of nature as we go from everyday scales to the edge of the universe. In a similar way, we should expect the description at Planck scale to be very different from what we are accustomed to in everyday life.

It is likely that, at Planck scales, space and time themselves might become discrete in a manner similar to the iron rod being described by discrete atoms at the microscopic scales. So, one is looking for some kind of 'atoms of spacetime' for providing the microscopic description of gravity and the familiar description in terms of Einstein's theory, curved spacetime etc. is expected to arise as an approximation at large scales. This is exactly in analogy with the manner in which elasticity arises at scales much larger than inter-atomic spacing.

The difficulty of course is that no one has succeeded in obtaining a sensible model for the 'atoms of spacetime'. The task is quite difficult and is similar to the demand for arriving at the atomic scale description of matter, given only the elastic properties of matter! In arriving at the quantum theory of matter, we had a host of experimental results to guide us while in the case of quantum gravity we are virtually proceeding in a blind manner. Given this situation, it is worthwhile to explore areas in which quantum mechanics has to coexist with strong gravity. This is what we will be concerned with for most part of the rest of the chapter.

6.2   Particles from the vacuum

Given the fact that developing a full quantum theory of gravity is still a distant dream, people have devoted a significant amount of effort on understanding the simpler situation in which one has to combine the principles of quantum theory and gravity without actually having to quantize gravity. This, subject usually called quantum field theory in curved spacetime, has lead to several beautiful results in the past and still remains a frontier research area.

What does it mean to say that one is studying a quantum field theory in an external gravitational field? Why should such a study lead to anything worthwhile? To answer these questions, it is again useful to start with a simpler case involving electromagnetism. We saw in Chapter 4 that the interaction of charged particles with the electromagnetic field is quite well understood by the theory which we call quantum electrodynamics. This theory treats both the charged particles — which, for the sake of definiteness, we shall take to be made of electrons and positrons — as well as the quanta of electromagnetic field — viz., the photons — as fully quantum mechanical. As we described in Chapter 4, the success of quantum electrodynamics is largely due to the fact that one can calculate the amplitudes for different processes in a perturbation theory, which is essentially like summing up a series of terms like $1 + x + x^2$.. etc. where x is a small number. One can get better and better accuracy for the final result by retaining a larger and larger number of terms in this series. Further, since x is a small number, you can also stop the summation at any stage and be confident that the terms which have been ignored are quite small compared to those which have been retained.

There are, however, some very interesting quantum processes involving the electromagnetic field and charged particles which *cannot* be obtained by this procedure. One of them is called the Schwinger effect, named after Julian Schwinger who was one of the creators of quantum electrodynamics

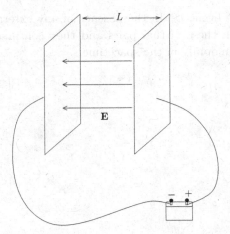

Fig. 6.2 A simple mechanism to generate a constant uniform electric field in a region of space. Two parallel conducting plates, kept separated by a distance L, and charged using a battery will host a uniform electric field in the region between the plates.

and received a Nobel Prize for the same. In simplest terms, this effect can be stated as follows. Consider a region of space in which there exists a constant, uniform electric field. One simple way of achieving this is shown in Fig. 6.2. We set up two large parallel conducting plates separated by some distance L and connect them to the opposite poles of a battery. This charges the plates and produces a constant electric field between them. Schwinger showed that, in such a configuration, electrons and positrons will spontaneously appear in the region between the plates through a process which is called pair production from the vacuum. This terminology is based on the fact that the particles are produced out of what we would have considered as the vacuum state. This is an important and somewhat counter-intuitive phenomenon which deserves a detailed description.

The first question one would ask is how particles can appear out of nowhere. This is natural since we haven't seen tennis balls or chairs appear out of the vacuum spontaneously. But the way we described the quantum field theory in Chapter 4 should make it less of a surprise. We pointed out that what we call vacuum is actually bristling with quantum fluctuations of the fields which can be interpreted in terms of virtual particle-antiparticle pairs. Under normal circumstances, such a virtual electron-positron pair will be described by the situation in the left frame of Fig. 6.3. We think of an electron and positron being created at the event A and then getting annihilated at the event B. In the absence of any external fields, there is no force acting on these virtual pairs and they continuously appear and disappear quite randomly in the spacetime.

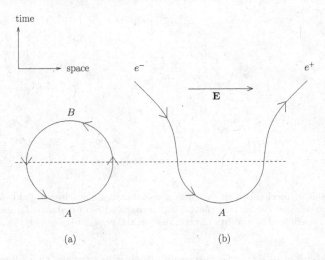

Fig. 6.3   In the vacuum there will exist virtual electron-positron pairs which are constantly created and annihilated as shown in the left frame (a). An electron-positron pair is created at A and annihilated at B with the positron being interpreted as an electron going backward in time. The right frame (b) shows how, in the presence of an electric field, this virtual process can lead to creation of real electrons and positrons.

Consider now what happens if there is a electric field present in this region of space. The electric field will pull the electron in one direction and push the positron in the opposite direction since the electrons and positrons carry opposite charges. In the process the electric field will do work on the virtual particle-antiparticle pair and hence will supply energy to them. If the field is strong enough, it can supply an energy greater than the rest energy of the two charged particles which is just $2 \times mc^2$ where m is the mass of the particle. This allows the virtual particles to become real. That is how the constant electric field between two conducting parallel plates produces particles out of the vacuum. It essentially does work on the virtual electron-positron pairs which are present in the spacetime and converts them into real particles as shown in the right frame of Fig. 6.3(b).

In fact, the only reason we find the phenomenon of particles appearing out of nowhere to be bizarre is because we would like energy to be conserved. Real particles have real energy while vacuum is assumed to have no energy.

So someone has to supply the energy for such a process to occur and, in our case, the electric field does this. The particles which are produced will move towards the two plates and will reduce the amount of charge which is present in them. This, in turn, will reduce the electric field and thus — if no corrective steps are taken — the process will come to an end on its own. To maintain a constant electric field we need to supply energy which, in this case, is done by the external battery source. It is essentially the electromagnetic field energy present between the capacitor plates which manifests as the energy of the produced particle.

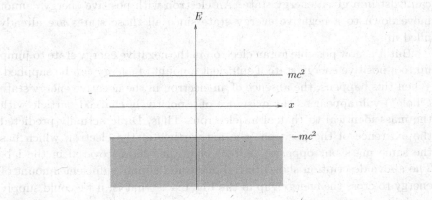

Fig. 6.4 Schematic description of energy levels allowed for an electron in Dirac's description. In addition to positive energies larger than mc^2 — which will describe a moving electron with rest energy plus kinetic energy — one can also have negative energy states with energy less than $-mc^2$. To prevent the electron from cascading down to these negative energy states, Dirac postulated that all these states (in the shaded area) are filled up.

There is a very picturesque way of describing this phenomena which is quite useful. This is based on a description of quantum electrodynamics originally due to Dirac. Dirac was the first person to write down an equation describing the field associated with an electron. This is similar in spirit to the Maxwell's equations which can be thought of as describing the field associated with a photon (which, of course, is the electromagnetic field). When Dirac solved his equation he found that something peculiar happened. The solution allowed for both positive and negative energy states, as shown symbolically in Fig. 6.4. All the states above the line $E = mc^2$, of course, describe the electron which has the rest energy mc^2. By adding to the rest energy some amount of kinetic energy, one can get to the energy states above mc^2. But there are also equally valid energy levels starting from the negative energy $E = -mc^2$ and going all the way downwards. This is "dangerous" because nothing prevents an electron with a positive energy to sink down to a negative energy state spontaneously and keep going down in energy; such a situation would be unstable and we do see electrons remaining as stable particles. Dirac solved this problem by a fairly brilliant stroke. (It is brilliant because he got the right answer by completely wrong arguments!). He postulated that all the negative energy states are already occupied by electrons. Since these are Fermi particles, only one particle can exist in a given energy state. An electron with positive energy cannot move down to a negative energy state since all those states are already filled up.

But it is now possible for an electron in the negative energy state to jump up to a positive energy state if sufficient amount of energy can be supplied. When this happens, the absence of an electron in the negative energy state ("hole") will appear as the existence of a positively charged particle with the mass identical to that of an electron. Thus, Dirac actually predicted the existence of the positron (the antiparticle of the electron which has the same mass but opposite charge) before it was discovered in the lab. The above description shows that if one could supply sufficient amount of energy to cross the energy gap in the Fig. 6.4 — that is, if we could supply a minimum energy $2mc^2$ — one can produce simultaneously an electron and a positron out of the vacuum. As we described before, this is precisely what happened in the case of Schwinger effect.

One can describe the action of an electric field in Dirac's language very easily. This is shown in Fig. 6.5 which describes the modification of the energy levels of electrons and positrons in the presence of an external electric field. The electric field is associated with a certain amount of potential energy which changes the energy levels from the ones shown in Fig. 6.4 (in the absence of the electric field) to the ones shown in Fig. 6.5. It is now clear the an electron with negative energy can move along the line PQ and appear in the positive energy sector, leaving behind a "hole" which is a positron. The tilting of the energy levels in Fig. 6.5 demonstrates how a constant electric field can convert virtual particles in the vacuum into real particles.

The interesting feature about the Schwinger effect described above is that it cannot be obtained from standard quantum electrodynamics using perturbation theory. One says that this is a non-perturbative effect involving the electromagnetic field and charged particles. In fact, to the first

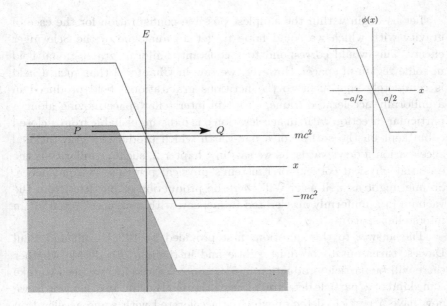

Fig. 6.5 The modification of Dirac energy levels in the presence of a strong electric field in some region of space. The electrostatic potential energy corresponding to the constant electric field is shown in the inset at top right.

approximation, the electric field is now treated as a purely classical, externally specified, quantity in the study of this problem. All that we did was to look at the quantum theory of the field corresponding to the electron and positron in an external electric field which was kept classical. We did not, for example, try to interpret the constant electric field produced by the two conductor plates in terms of photons which are the quanta of the field. In fact, it turns out that it is quite complicated to do this in a correct manner. So the Schwinger effect can be thought of as a result in quantum field theory in an external, classical electromagnetic field.

Move over from the electromagnetic field to the gravitational field. We could ask whether there is an analog for the Schwinger effect in the case of gravitational field. Will there be production of particle-antiparticle pairs in an externally specified classical gravitational field? Such questions would be worth studying since we do not have to treat the gravitational field as built out of gravitons in this context and still get an insight into processes that involve both gravity and quantum theory. In fact the results of such investigations have been quite surprising, to say the least.

Let us begin within the simplest possible configuration for the case of gravity with which we could hope to get an analogue of the Schwinger effect. This would correspond to a constant, uniform, gravitational field in some region of space. However, we saw in Chapter 2 that such a field is completely equivalent to the fictitious gravitational field produced in a uniformly accelerated frame. A lift in interstellar space moving along a particular direction with an acceleration g is indistinguishable from a closed cabin kept on the surface of a flat Earth which produces a gravitational acceleration g downwards; as we saw in Chapter 2, such a similarity is the essential physical content of Einstein's physical principle of equivalence. So one might as well ask: Will there be production of particles from the vacuum in a uniformly accelerated frame, say, within an accelerated lift in interstellar space?

The answer to this question, first provided by Bill Unruh and Paul Davies, turns out to be quite subtle and intriguing. To decide whether there will be particle production inside an accelerated lift, we need to take some kind of a 'particle detector' along with us in the lift. In other words, we now have a particle detector which is accelerated with some acceleration g. It does not really matter what this particle detector is. We will just assume that it is some contraption which will click if a particle is around. For example, it could be an atomic system (with two energy levels, say) which is capable of absorbing a photon and moving from the ground state to the excited state. If this system gets excited, we could say that it has detected a photon. So, in practical terms, we can rephrase our original question as follows: Suppose we build a simple detector which will click when it is at rest if a particle hits it. We now put this detector inside a lift which is moving with an acceleration g. Will the detector click?

The answer turns out to be "yes". An accelerated detector will detect particles in a quantum state which we originally thought was a vacuum and hence is devoid of all particles. What is more intriguing is the following fact. A detector moving with acceleration g will behave like another detector which is static and is immersed in a box, the walls of which are kept at a temperature $T = (\hbar/ck_B)(g/2\pi)$. We saw in Chapter 1 that a system in a temperature T has a spectrum of photons which is characteristic of that temperature (see Fig. 1.16). A detector immersed in such a box, the walls of which are kept at a temperature T, will be able to measure this spectrum and thus determine the temperature. So, essentially the detector is acting as a thermometer in this context and is measuring the temperature of the gas of photons inside the box. What our result tells you is that if we accelerate such a thermometer in empty space, it will behave as though it is immersed in a gas of particles with a characteristic temperature that is proportional to the acceleration. Put more dramatically, the vacuum has a temperature which depends on who is measuring it. A person moving with a given acceleration will attribute a particular nonzero temperature to the vacuum while all inertial observers will swear that the vacuum has zero temperature!

You can already see that we have obtained a bizarre and revolutionary result. One interpretation of this result will be the following. A uniformly accelerated frame is indistinguishable from a constant gravitational field. So, a thermometer detecting a spectrum of particles with a given temperature in a uniformly accelerated frame can be thought of as essentially confirming particle production by a uniform gravitational field. That would be in perfect analogy with the Schwinger effect.

The trouble with this interpretation is the following: Since the constant gravitational field is completely indistinguishable from an accelerated frame, one can "switch-off" such a field by going to another coordinate frame which, in this case, is just an inertial frame. An observer in the inertial frame will see no particles while another observer accelerating nearby will claim there is a thermal spectrum of particles with a nonzero temperature. So you reach the inevitable conclusion that the concept of particle depends on the observer. What one person calls a vacuum state, another might call a many particle state.

This sort of thing has happened before in physics, the classic example being the notion of flow of time. In non-relativistic physics we assume that the rate of flow of time measured by clocks is an absolute quantity and is independent of the state of motion of the observer carrying the clocks. When one second elapses for a given observer, all other observers in any state of motion with respect to the first one, will also record the lapse of one second — which is what we thought. With the advent of relativity, we knew that this is not true and that the flow of time is not absolute but relative. Each observer has his own notion of time and the best we can do is to give the precise relationships between the notions of time of different observers.

In a similar manner we had assumed that the notion of a particle and that of a vacuum state are absolute concepts in special relativistic quantum field theory. This is the sort of structure which is normally used in particle physics to describe, say, quantum electrodynamics. Such a framework, which we described in Chapter 4 incorporates the principles of quantum mechanics and *special* relativity but does not take into account the effects of *general* relativity. It is constructed in such a way that all inertial observers will agree with the notion of particles. That is, if an observer detects one photon in a given state of a system, any other observer who is moving with respect to the first one with a uniform velocity will also claim that the state has one photon. Similarly, if the first observer declares a state to be a vacuum (that is, no photons) then all other observers moving with respect to the first one with a uniform velocity will agree that it is a vacuum state. This is built into the standard quantum field theory and hence one assumes that the notion of particles or a vacuum state is observer independent.

It is this notion which is shattered when one compares two observers who are moving with respect to each other with a relative acceleration. When one observer declares a state as a vacuum state, another observer who is moving with an acceleration will declare it as a state filled with particles; in fact, the second observer will call it a thermal state corresponding to a specific temperature. So the concept of particle is relative and observer dependent when we take into account accelerated observers.

A consequence of this result is that one cannot easily answer the question: "Are the particles seen by the accelerated detector real?" We have a situation in which one observer declares a state as devoid of particles while another (accelerated) observer thinks of the same state as made of many particles. The real lesson to draw from this phenomenon is that one can determine the reality of a physical process only with respect to the procedure which is used to determine it. If you take a particle detector and run with it, it will certainly click; in the simple two-energy-state model of the detector, it would have made a transition from the ground state to the excited state absorbing some energy. This is, of course, as real as any process can get. We can further ask where the energy came from which made the detector get excited. The energy has to be supplied by the source which is accelerating the detector. To keep a particle detector moving with a constant acceleration, you have to supply somewhat more energy than if it were not designed to be a particle detector. This energy goes into exciting the detector and there is no fundamental contradiction. But at the same time, the observer moving with the detector will claim that the particles and the energy came from the environment in which there is a thermal distribution of particles.

Though we started our discussion by comparing the constant gravitational field with the constant electric field, one can now see that there is one crucial difference. When an electric field produces pairs of particles, all observers agree on that phenomena. But as we have seen, in the case of a uniform gravitational field the interpretation is different for the two observers.

There are other situations in which the gravitational particle production is more like the case of the uniform electric field. In the presence of a genuine gravitational field, we know that the spacetime is curved and one can use any suitable set of coordinates to describe it. In such a context, the concept of particles becomes fairly muddled and is dependent somewhat loosely on the concept of the observer that is used to define the particles. Roughly speaking, each class of observers will define their own notion of particles and the best we can do is to provide the rules that will allow us to relate these different particle concepts. It is exactly analogous to what happened to the notion of time in special relativity.

That, however, does not explain where the *temperature* came from. It is one thing to say that two different observers in a relative accelerated motion will differ about their notion of particles and one observer will see particles in a state declared to be a vacuum by the other; it is a different and more curious thing to say that the vacuum state will appear to be a thermal state for the uniformly accelerated observer. As we said before, thermal states are very special and require the distribution of energy among the particles to have a particular form (as shown in Fig. 1.16). Why should the vacuum state appear to be a thermal state in this context?

Before we answer this question — which we shall postpone to the next section — it is worth pointing out that not all gravitational fields produce particles which have a thermal spectrum. An important example in this context is provided by the expanding universe itself. If one studies quantum field theory in the background of an expanding universe, one reaches a conclusion similar to the one obtained earlier. The gravitational field of the expanding universe will produce particles out of the vacuum. However, in general, the spectrum of the particles which are produced will not be thermal and one cannot attribute a temperature to it. This shows that, while the idea of particles being produced by an external gravitational field is of general validity, one being able to associate a temperature with them is a peculiar feature which arises only in some special cases. But these are the cases which lead to a fairly deep insight about the nature of gravity and we shall explore them in the next section.

6.3   Horizon, entropy and temperature

While discussing the nature of heat in Chapter 1, we pointed out that some thermodynamic concepts like temperature, entropy etc. are purely macroscopic and arise in systems with a large number of degrees of freedom. It is therefore very puzzling that an accelerated observer will attribute a temperature to the vacuum state itself. A closer examination of this concept, which we shall now describe, leads to an interesting connection between gravity and thermodynamics.

To explore this connection, we first have to describe in detail the notion of a horizon. In everyday parlance, the horizon is essentially as far as you can see from a tall building, say. The idea in the case of gravity is similar. Consider a black hole of a given mass located around the origin and let us assume that you are staying at a safe distance from the black hole. We have already seen in Chapter 2 that no signal, not even light, can escape from the inside of the black hole and reach you. The size of the black hole which defines this region from which signals cannot reach the external world, is a sphere of radius $2GM/c^2$ where M is the mass of the black hole. When you look at the black hole from the outside, the same radius defines the horizon which limits your visibility. You cannot see anything beyond that radius which is at the interior of the black hole. Hence this surface is called the *event horizon* of a black hole.

What can come as a surprise to you is that horizons can arise in very many other situations in general relativity. The fundamental reason for the existence of the horizon is the influence of the gravitational field on light rays. One can always arrange matters such that there are classes of observers in a given spacetime who cannot receive information from some region in the spacetime. This is illustrated in Fig. 6.6. Consider an observer following the trajectory like the ones shown in Fig. 6.6 which asymptotically approach a 45 degree line passing through the origin in a spacetime diagram. It is now obvious that no signal can ever go from the upper left hand part of the diagram to the observer following the trajectory shown in the figure. This entire region beyond the line marked horizon is inaccessible to this particular observer. This kind of a situation occurs in several contexts in general relativity. When there is a black hole centered around the origin, the trajectory of an observer who is staying put outside the black hole will be given exactly by a curve like the one in Fig. 6.6. A still simpler example which we want to focus on corresponds to a uniformly accelerated observer in flat spacetime. Remarkably enough, this situation is also described by Fig. 6.6. The trajectory now corresponds to an observer who is travelling with a uniform acceleration at late times and approaching the speed of light asymptotically. Such an observer will find that just because of the trajectory, she is completely cut-off from the region beyond the horizon. Mathematically, the two situations — an observer who is uniformly accelerated in flat spacetime and an observer who is at rest outside a black hole — map to the same figure as far as the phenomena we are interested in are concerned. (Of course, these two situations differ in many other details which are, fortunately, irrelevant to our discussions.)

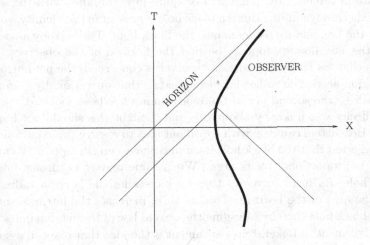

Fig. 6.6 The concept of horizon in gravitational theories. The thick line denotes the trajectory of a given observer. The light rays travel at 45 degrees in the coordinate system shown here. It is obvious that no signal can reach the observer from the region above the line marked horizon.

One can also see from this discussion that the existence of the horizon is an observer dependent phenomenon. Someone who is staying outside the black hole has her vision blocked by the horizon. But if the same observer decides to take a plunge into the black hole, she will cross the horizon and will acquire more information than an observer who is confined to the outside. In the case of an accelerated observer, she only has to stop accelerating and all will be revealed to her; there will no longer be any horizon. Hence, in both the cases, the existence of the horizon in the practical sense of limiting one's vision is intimately linked to the state of motion of the observer.

To understand what horizons have to do with the temperature perceived by the accelerated observer, we shall take a historical approach and first discuss the situation in the case of a black hole. The story should probably begin with the conversation which took place between John Wheeler and his then graduate student Jacob Bekenstein in 1970. John Wheeler was intrigued by the implications of black hole horizons for the second law of thermodynamics which we described briefly in Chapter 1, Sec. 1.5. The cause for this concern can be described as follows. Let us say that some physical process takes place in a small box which causes some amount of increase in the entropy. That will be quite in accordance with the second law of thermodynamics. But if a black hole is present in the vicinity, you can throw the box and its contents into the black hole. The entropy associated with the box now disappears beyond the horizon of the observer. This means that, as far as the outside observer is concerned, the net entropy in the region accessible to her has decreased so that operationally — for this observer — the second law of thermodynamics has been violated.

Wheeler was unhappy about this and thought this should not happen. Jacob Bekenstein came up with a brilliant idea to resolve this contradiction. He suggested that the black hole itself must have certain amount of entropy which is proportional to its area. When some matter is thrown into the black hole, its mass increases thereby increasing the horizon radius and thus the area of the horizon. This, in turn, increases the intrinsic entropy of the black hole thereby salvaging the second law of thermodynamics. The crucial element in Bekenstein's argument is the idea that one can associate an entropy to the black hole which is proportional to its area. A few years earlier Stephen Hawking had shown that the areas of black holes do have one property which makes this identification possible. He had shown that, in any interaction involving several black holes like — for example, the merging of two black holes leading to the formation of a bigger single black hole — the sum of the areas of the black holes cannot decrease. This is very similar to what happens to entropy in normal physical processes. As we described in Sec. 1.5, the sum of the entropies of all the components never decreases during a physical process. Bekenstein used this result of Hawking to conjecture that the entropy of a black hole is proportional to its area.

Unfortunately, the rest of the community — including Hawking — did not immediately agree with this notion. Their argument went as follows. Black holes have a mass and hence an energy Mc^2. If we now attribute an entropy to the black hole which is proportional to its area, then one can work out a resulting temperature for the black hole. This result arises from basic principles of thermodynamics which gives a procedure for computing the temperature for any system for which the entropy can be given as a function of its energy. The key point is that the temperature of a black hole has to be nonzero if it has a nonzero entropy. But then, if the black hole has a nonzero temperature, it will have to emit radiation with a Planck spectrum corresponding to this temperature. Hawking and others felt, in 1972, that this is impossible because nothing can escape from a black hole's event horizon. Arguing backwards they pointed out that since black holes cannot radiate anything they cannot have a temperature and hence they cannot have an entropy. Hawking wanted to study this situation closely and investigated the behaviour of quantum fields around a black hole. Interestingly enough, this analysis led to an intriguing result that the black holes indeed have a temperature and an entropy exactly as conjectured by Bekenstein. In other words, the horizon of the black hole produces particles and radiates them to infinity just like a hot stove kept at a given temperature. This is probably the single most important result which we have connecting the principles of quantum theory and general relativity and hence deserves a close scrutiny.

The simplest interpretation of this result will be as particle production in an external gravitational field. We saw earlier that in the presence of a strong electric field, the vacuum can destabilize itself and lead to the production of particle-antiparticle pairs. In Dirac's interpretation, the particles with negative energy are re-interpreted as antiparticles of positive energy. In the case of a black hole, we can arrange the negative energy particle produced due to the vacuum breakdown to fall inside the black hole and let the positive energy particle escape to infinity. This could explain why the black hole is radiating particles. When the negative energy particles fall into the black hole they will, of course, decrease the energy (that is, the mass) of the black hole. Thus, the radiation discovered by Hawking allows the mass of the black hole to be slowly converted into energy of the particles which are flowing to infinity. One usually calls this black hole evaporation because the phenomena is akin to molecules of a liquid evaporating from near the surface of the liquid thereby decreasing the energy content of the liquid.

This interpretation is fine as far as it goes but it leaves one crucial question unanswered. How come the spectrum of particles has a thermal distribution so that one can associate a temperature with this phenomena? When we studied the Schwinger effect, we again saw that an electric field can produce particles but we did not associate either a temperature or an entropy with the electric field. We clearly have a deeper message from nature in store in this case.

This is, of course, the second time we are coming across the notion of temperature in the context of gravity. The first case was discussed in the last section in which we saw that a detector accelerated through the vacuum will interpret the vacuum state as populated by a thermal distribution of particles. Historically, Hawking's analysis of the black hole and its temperature came first. Very soon, two other physicists Paul Davies and Bill Unruh independently pointed out that the concept of temperature arises even in the case of a uniformly accelerated detector or observer. (There is, however, a technical difference between the two results. In the case of an accelerated observer, she interprets the vacuum state as a many-particle state but it is not the breakdown of the vacuum which leads to the particles she sees. Any real energy absorbed by the detector comes from the person who is pushing the detector so as to keep it moving with a uniform acceleration. In the case of a black hole formed by the collapse of gravitating matter, the energy carried away from the particles to infinity comes from the mass of the black hole. In that sense, one might think that black hole evaporation is "more real" but this technical issue is irrelevant for our further discussion.) One clear common feature between the two cases is, of course, the existence of the horizon which suggests that somehow the horizon is the culprit in endowing certain gravitational fields with thermal properties.

The original idea of Bekenstein in assigning entropy to the horizon certainly fits with this interpretation. To see this, let us go back to the idea of entropy for normal systems (like a gas) which we discussed in Chapter 1. We saw there that the entropy is associated with lack of information about the microscopic states of the system. When we describe a gas made of zillions of molecules using just a handful of variables like pressure, volume and temperature, we are certainly losing a lot of information about the microscopic states which need to be described by giving the position and speed of each of the molecules. We pointed out in Chapter 1 that if there are 2^S different microscopic states which correspond to a given macroscopic state, then S is the entropy of that macroscopic state. In the case of a black hole, there is a curious fact which allows us to make a similar connection.

It was known for a long time that the black hole solutions in Einstein's theory can be described by very few parameters. Essentially we only need to specify the mass M, the electric charge Q and the angular momentum J

to describe the black hole. So, if we start with a star which has a particular shape, composition etc. and collapse it to form a black hole, all the extra information except those in M, Q and J gets lost. For example, if we start with two stars with zero electric charge ($Q = 0$) and same mass and angular momentum but completely different shapes and collapse them to form two black holes, they will be identical! Before they collapse to form a black hole, one can explore the gravitational field around them and determine, for example, their shape etc. Once they have become black holes, we can no longer do this, which clearly indicates loss of information. We can, therefore, take the point of view that the macroscopic parameters of the black hole (M, Q and J) describes its macroscopic state just as a handful of parameters like pressure, volume etc. describes the macroscopic state of a gas. There could be several microscopic possibilities which — as we have seen — can lead to the same macroscopic black hole configuration. If we can compute the total number of microscopic states from which a black hole with a given set of parameters M, J and Q can be formed, we can possibly compute its entropy. As we said before, once we know the entropy and the energy of the black hole which, in the simplest case, is just Mc^2, we can compute its temperature.

While this is a nice idea, it has so far been impossible to convert it into a rigorous mathematical formulation and actually compute the total number of states for a black hole with parameters M, Q and J. Classically, one can have an infinite number of possibilities by which one could have obtained a black hole of a given final configuration. This is because, classically, one can sub-divide matter to infinitesimally small parts so that any finite mass black hole can be obtained from an arbitrarily large number of sufficiently small particles. So, if we do the computation using classical general relativity, we are bound to get an infinitely large value for the entropy of the black hole.

We, of course, know the final answer that needs to be obtained. Bekenstein had originally said that the entropy is proportional to the black hole area but could not rigorously determine the proportionality constant. After Hawking's derivation of black hole evaporation, it became possible to do this and the result can be expressed as the equation $S = (1/4)(A/L_P^2)$ where A is the area of the black hole horizon and $L_P^2 = G\hbar/c^3$ is the fundamental constant in quantum gravity which we introduced in Sec. 6.1. Since L_P is the only constant which is relevant to the problem and S should be a dimensionless number, it is quite obvious that S must be proportional to A/L_P^2. The existence of the Planck constant $\hbar$ shows that the finite value of the entropy for the black hole is indeed a quantum phenomenon.

If we set $\hbar$ to zero, the Planck length L_P also vanishes and the entropy becomes infinitely large because we are dividing a finite quantity, viz. the area A, by zero. This is, of course, the result we originally found in classical general relativity because any black hole of finite mass can be made out of an arbitrarily large number of particles of sufficiently small mass. This is not possible in quantum theory. A particle of mass m in quantum theory cannot be confined to a size smaller than $\hbar/mc$, called the Compton wavelength of the particle. One way of understanding this result is to go back to the uncertainty principle which says that the uncertainty in the position ϵ has to be larger than $\hbar/mv$ where v is the uncertainty in the velocity. If we assume that v has to be less than speed of light, c, it follows that the uncertainty in the position of a particle of mass m has to be larger than $\hbar/mc$. [This argument is not rigorously valid but the final result is!] So particles with smaller and smaller mass will have larger and larger values for $\hbar/mc$. So if we try to build a black hole with particles with very tiny mass, its Compton wavelength will be so large that we can no longer be sure that the particle is actually inside the black hole! This means, in quantum theory, there is a lower limit to the mass of the particle which can be used to build a black hole of a given size. Hence we get a finite value for the entropy when we bring in the quantum nature of matter. In fact, it is fairly elementary to use this argument to prove that the entropy has to be proportional to A/L_P^2 but this argument cannot determine the proportionality constant. What Hawking's calculations show is that the coefficient in front must be $(1/4)$ which has become a holy grail for theoreticians to try and obtain.

Since we know the correct expression for the entropy, we also know that the number of microscopic states of a black hole with horizon area A is equal to $2^S = 2^{(1/4)(A/L_P^2)}$. This can be given a fairly interesting interpretation qualitatively. The tiny lengthscale in quantum gravity, L_P is the scale at which quantum gravitational effects will be important. Work by Padmanabhan and many others have shown that it is not possible to devise measurements which could measure lengthscales smaller than L_P. Roughly speaking, to devise an instrument which can probe such small scales, you need to put too much of mass in too tiny a region that the instrument itself will collapse and form a black hole. This shows that our ideas of spacetime continuum breaks down at scales below Planck length. This is somewhat similar to the breakdown of the notion of a continuous solid at scales close to the inter-atomic separation or the lattice spacing of a crystal lattice. Taking a cue from this, we can speculate that the spacetime probably has a discretized structure at Planck scales. The surface of the black hole horizon can now be divided into small area elements, each of area $4L_P^2$, say. This means there are $S = A/4L_P^2$ discrete elements on the surface of the horizon. If we endow each one of them with one bit of information by assigning, say, a number 0 or 1 to each one of them, then the total possible ways this information can be coded is also $2 \times 2 \times 2 ..... S$ times $= 2^S$ which will, of course, correspond to an entropy S.

This argument, unfortunately, cannot be taken too seriously because we have put in couple of crucial facts by hand. First, we could have said that the lattice spacing — or the areas of each element — is L_P^2 rather than $4L_P^2$; there is no reason to choose the number 4 in that argument except to obtain the known final result. Different choices will give different answers of which only one is correct. One could have also assumed that each of the area elements can exist in $3, 4, ...$ possible states rather than just 2 possible states. Again this will change the proportionality factor between the area and the entropy. In that sense, this argument does not tell us anything more than what one could have obtained from simple dimensional considerations.

The true value of the above approach, however, lies elsewhere. The really important question, in some sense, is the following. What are the microscopic degrees of freedom which contribute to the entropy of the horizon? We know that the entropy of a gas comes from ignoring the degrees of freedom associated with the atoms of the gas. What are the analogous atoms of spacetime? Our first guess was that this should correspond to the information contained in the matter that has collapsed to form the black hole. But if this is the case, it is not clear whether the entropy of the black hole should increase as the area or as the volume. The result $S \propto A$ strongly suggests that the relevant degrees of freedom lie at the surface of the horizon rather than in the interior of the black hole. This does not quite exclude the possibility that the entropy is somehow related to the different configurations of matter which form the black hole. The matter that collapses actually takes infinite time to reach the horizon surface as far as an outside observer is concerned. So it can somehow happen that the area of the horizon determines the entropy rather than the bulk. But it looks somewhat unnatural. On the other hand, no one has really come up with a viable set of degrees of freedom which could lead to the correct entropy.

Two leading contenders which make claims for some partial success in this problem are the string theory and loop quantum gravity. These two represent elaborate mathematical structures which might have ingredients that might help to unify the principles of quantum theory and general relativity. In spite of several serious differences which exist between these two approaches, they both have one thing in common. Both of them incorporate the Planck length as some kind of minimal length in spacetime and as such they possess, in principle, a possible mechanism for obtaining finite number of microstates in a black hole. Both these approaches have achieved a partial success but the situation is far from being completely satisfactory.

In the case of string theory, one can perform the calculation rigorously only for a very special class of black holes called extremal black holes. These are a very peculiar kind of black holes with both charge and mass and having such a large amount of charge that the gravitational field produced by the mass is comparable to the gravitational field produced by the electric field due to the charge. In classical general relativity, such solutions are considered as a curiosity since it is unlikely that any real black hole which forms in the universe will have such a large amount of charge. What is more, the Hawking temperature for such extremal black holes happens to be zero so that they do not radiate at all. They nevertheless have entropy since the horizon has a finite area. Using the mathematics of string theory, it is possible to map such an extremal black hole solution to another field theory for which one can compute the number of microstates. One can also provide an argument as to why the number of microstates for this theory should be essentially the same as that for the original extremal black hole solution. This calculation does lead to the proportionality factor $(1/4)$ between the entropy and (A/L_P^2) which many string theorists claim to be a success. But as we said before, this result is rigorously valid only for a very special class of black holes which are pretty much unrealistic. The string theory, for example, cannot be used to determine the entropy of the simplest kind of black hole solutions like the one in which the gravitational field is determined by the mass of the black hole without any electric charge.

The loop quantum gravity, on the other hand, can be used to compute the entropy of almost any kind of black hole but they cannot obtain the factor $(1/4)$. This theory has an undetermined parameter in it and it turns out that the final answer for the entropy depends on this undetermined parameter. If we know the correct result, then one can choose the value of the parameter so as to be consistent with it but this cannot be considered as a derivation from first principles. After all, the proportionality between entropy and A/L_P^2 can be obtained from half-a-dozen different arguments and hence it is the proportionality constant which is crucial. Practitioners of this art stress the success in handling realistic black holes which string theorists cannot do. On the other hand, string theorists point out that, at least for one kind of black holes, they can get the factor (1/4) which loop quantum gravity cannot obtain. It is fair to say that neither approach has led to a satisfactory answer in spite of all these claims.

6.4 Gravity as an emergent phenomenon

While we still do not have a satisfactory theory of quantum gravity, the ideas which have emerged from the study of horizons and the connection between thermodynamics and gravity provide some insights into quantum gravity. These insights are largely independent of the microscopic descriptions of quantum gravity which may emerge at a future date and hence are likely to survive the changes in our ideas of what quantum gravity should be. Essentially, the results of the previous section suggest a paradigm in which we interpret gravity as an emergent phenomenon, like elasticity, and provide an effective description for it which is largely independent of the microscopic theory.

To appreciate this paradigm, let us again consider the description of a solid at different levels. An engineer, who has to use an iron rod to bolster a structure, hardly worries about the microscopic equations governing the atoms in the iron rod. She will think of the iron rod as a continuum with some density and elastic properties. The latter is described in terms of a set of numbers called elastic constants which will describe how the iron rod responds to stresses and strains. One of these numbers, for example, will tell the engineer how much the iron rod of a certain dimension will bend if a given weight is attached to one of its ends. Given such information, the engineer can describe the behaviour of the iron rod without ever having to wonder what it is made of.

At the same time, we know that the iron rod can be heated, which implies that it has a mechanism to store the heat energy which we supply internally. Boltzmann was the first person to realize that this indicates the existence of some internal degrees of freedom. Without this microscopic structure — in the continuum approximation in which the iron rod is described by a smooth interior throughout — one cannot heat up the iron rod and talk about its temperature. When the temperature comes into play, we have to move from a description based on a continuum and the elastic constants to a thermodynamic description involving the temperature, entropy and laws of thermodynamics. As we saw in Chapter 1, it is in principle possible to obtain the laws of thermodynamics from the microscopic description but the final result is largely independent of the actual microscopic structure that is assumed. For example, an ideal gas of helium or an ideal gas of argon will obey the same gas equation $PV = Nk_BT$ even though the atomic structure of helium is very different from the atomic structure of argon. The details of the microstructure do not affect the thermodynamic description but the existence of the microstructure is absolutely vital.

Let us now consider gravity along the same lines as some kind of elasticity of spacetime. This idea was originally due to Russian physicist Sakharov but, of course, the details of the model need to be changed incorporating what we know about the thermodynamics of horizons in recent years. Such a formalism was developed in the last several years by Padmanabhan and others. This approach provides an exact parallel between what we described above for an iron rod and our description of spacetime. At macroscopic scales, the space and time appears to be a continuum just as a solid appears to be a continuum at scales much bigger than the lattice spacing of atoms. Just as the solid is governed by laws of elasticity, we can describe the spacetime by Einstein's equations. Just as a solid can exist at a nonzero temperature, the spacetime can also exhibit a nonzero temperature. We saw in Sec. 6.3 that a spacetime need not be special for it to have a temperature with respect to a given observer. If the trajectory of an observer is such that she has a horizon blocking a certain amount of information, then she will perceive the spacetime to be hot. Black hole solutions and the like come with a more natural concept of horizon but even here the horizon is an observer dependent concept. For example, an observer who dares to plunge into the black hole will have access to more information than an observer who is orbiting safely outside the black hole. This fact will reflect on the amount of entropy attributed to the black hole by the suicidal observer. When one usually talks about the entropy of a black hole, we do have in mind a particular class of observers though we may not explicitly state this.

At this level, we have to describe a hot spacetime just as in thermodynamics we have to describe a hot solid. We are again forced to the notion, following Boltzmann, that the capacity of the spacetime to exhibit nonzero temperature implies the existence of microscopic degrees of freedom in the spacetime. These "atoms of spacetime" must be ultimately described in terms of a full quantum theory of gravitation. But just as the thermodynamic behaviour of an ideal gas is largely independent of the actual microscopic physics, we would expect the thermodynamics of spacetime to be largely independent of the microscopic structure of the theory. What is important is the existence of the micro-structure, not the specific details of the micro-structure.

If these ideas are correct, then it must be possible to reinterpret the dynamical equation governing the behaviour of spacetimes with horizon (like, for example, the black hole spacetimes) entirely in thermodynamic language. It turns out that this is indeed true. If you look at Einstein's equations describing a black hole near the horizon, it is possible to rewrite it as a first law of thermodynamics involving some amount of internal energy due to a finite temperature and the usual mechanical energy which arises from the normal pressure of the matter etc. Viewed from this angle, any horizon behaves like a thermodynamic system and the laws of thermodynamics reduce to what we have called Einstein's equations. This provides the connection between the microscopic and macroscopic physics just as the ideal gas law, for example, connects the macroscopic behaviour of the gas to the existence of microscopic molecular structure.

In fact, one can use these ideas to provide a completely alternative paradigm for gravity itself. To see this, let us go back to the historical sequence of events which led to attributing an entropy to black hole horizons. As the discussion between John Wheeler and Bekenstein revealed, one can hide entropy behind a horizon and violate the second law of thermodynamics unless the horizon itself has certain amount of entropy. Usually, one considers this process in the context of dropping some amount of matter into a black hole horizon. But the basic idea is far more general. If one considers a displacement of a horizon by a small amount (for example, by imagining that the radius of a black hole is increased slightly), the horizon will engulf some entropy of the matter surrounding it. This will violate the second law of thermodynamics unless such displacements of a horizon itself cost a certain amount of entropy. In fact, this result should hold for every observer in a spacetime in any state of motion. We have already seen that different classes of observers attribute to the spacetime different properties because they may have access to different regions of spacetime. Any one of them can think of a displacement of the horizon as perceived by them leading to a violation of the second law of thermodynamics. To ensure that the second law of thermodynamics holds for every observer in any state of motion, we need to make sure that an entropy is associated with the displacement of any horizon in a specific manner.

In this approach, we have two sources of entropy. One is related to the normal matter and the other is related to the displacement of horizons perceived by observers in different states of motion. We have seen in Chapter 1 that when any system comes into equilibrium, its entropy reaches a maximum. In this particular case, we need to maximize the total entropy as perceived by every possible observer. This condition should lead to the dynamics of the spacetime.

Remarkably enough, such a formalism leads to exactly the same equations as in Einstein's theory at the lowest order. What is more, one can write down the explicit corrections to the Einstein's theory in this formalism in a well defined manner. So the picture of gravity as an emergent phenomena like elasticity can be given a rigorous mathematical backing in the thermodynamic limit. Further, the final results are fairly independent of the detailed microscopic theory of quantum gravity.

The last point has one unfortunate consequence. While this approach is conceptually elegant and gives a new paradigm for describing gravity, it is not capable of telling us more about the microscopic structure. Of course, this is too much to ask for from such a paradigm. After all, the description of an ideal gas in terms of a large number of molecules moving at random, by Boltzmann, cannot be pursued further to actually reveal the atomic physics of the molecule. The fact that the thermodynamic description is independent of the microscopic details cuts both ways. It allows you to formulate the theory without worrying about the detailed microstructure but at the same time, refuses to reveal anything about the microscopic structure.

6.5   Black holes, singularity and the information

In the last few sections, we concentrated on the connection between the thermodynamics of the horizon and how it leads us to a new paradigm of gravity. In this section we shall again return to the study of black hole evaporation and some further conundrums it produces. One of the issues which has bothered researchers and has led to some amount of controversy is related to the loss of information which accompanies the formation of a black hole. But this problem is so intimately connected with several other properties of the black hole, that we first need to develop some more background material to understand these issues.

Consider a black hole which forms due to the collapse of a large amount of matter like, for example, at the end stage of stellar evolution. We will concentrate on a particle located at the surface of the star which is under-going the collapse or, just to make it a little bit more dramatic, we will consider a scientist Joe who is willing to collapse into a black hole along with the matter by staying on the surface of the star. Joe is in constant communication with his scientist colleague Moe who is located outside far away from the collapsing matter and is playing it safe. Joe and Moe can communicate with each other by light signals and Joe has agreed to send light signals every hour on the hour according to the clock he is carrying.

As the star collapses, Joe notices that he is moving towards the origin along with the stellar material and is dutifully sending signals to Moe at periodic intervals. He is also receiving back the communication from Moe. We know that eventually the star will collapse to a radius which is smaller than the size of its horizon and once this happens, Joe will miss his last chance to go back to the outside world. As we described in Chapter 2, this is like the light fish which gets trapped far too downstream in a fast moving river and cannot swim back upstream because the current is too strong. Of course, the light fish sees absolutely nothing peculiar at the point H (see Chapter 2, Sec. 2.5) which acts as a horizon when it is swimming downstream in blissful ignorance. Similarly, Joe will notice nothing peculiar as he crosses the horizon when the star is collapsing.

The situation from Moe's point of view is very different. To begin with, his clocks are located at a very moderate gravitational field compared to Joe's clocks which are in the vicinity of a very compact star that is collapsing to form a black hole. We saw in Chapter 2 that gravity affects the flow of time and slows down the clock. Near the horizon of the black hole, this effect is phenomenally large and the clocks almost come to a standstill. Its practical implication is the following. Even though Joe is sending light signals once every hour, Moe receives these signals with larger and larger intervals. The light signal Joe sends just before crossing the horizon takes infinite time to reach Moe. As a consequence, Moe never sees his friend plunge through the horizon! In fact, all the collapsing matter hovers just outside the horizon of the black hole as far as Moe is concerned. Joe, on the other hand, is sending light signals even after he has crossed the horizon but, of course, these signals never reach the outside world since they cannot escape from inside the horizon. We see that an extreme level of observer dependence has been brought into the description of events when a black hole forms.

What actually happens to Joe and all the stellar matter, as the collapse progresses? Classical general relativity gives a rather unsatisfactory answer.

First, it tells you that, operationally speaking, the black hole takes an infinitely large amount of time to form as far as the outside observer like Moe is concerned. All the matter in the star, as well as Joe, hovers around just outside the black hole horizon for an infinite time. The situation from Joe's point of view is a lot more grim. He keeps plunging to the center with all the matter being crushed to smaller and smaller radii in a finite time according to his clock. In other words, he is crushed to death by the infinitely large gravitational pull within a finite period of time. All the matter is condensed to a single point of zero radius and infinite density and, of course, infinite gravitational force. Such a situation is called a singularity in mathematics and physicists hate singularities. The problem with the singularity is that the equations of motion, like Einstein's equations, break down at the singularity and one cannot predict the future evolution. This, quite rightly, bothers physicists but classical relativity provides no answer.

If you have been following the arguments in the last two sections, you would of course have realized that classical gravity in any case is not adequate to describe black hole physics in its extreme form. There are two reasons for it. First, we have not taken into account the effects of black hole evaporation in the description given above. Second, when the matter is crushed to a size comparable to Planck length, quantum gravitational effects will become important and one simply cannot trust the classical picture which predicts zero radius and infinite density for the collapsing matter.

Of these two issues, the second one is still completely unsolved. Nobody knows what happens to the matter when the collapse proceeds to Planck size. We do not have today a viable quantum theory of gravity which can be used to provide an alternative description. We mentioned earlier that both string theory and loop quantum gravity have attempted to provide an explanation for black hole *entropy*. Neither of these approaches can tell us anything definite about what happens to the collapsing matter when it is compressed to sizes of the order of Planck length. This is yet another reason why one would like to be cautious in taking the claims for entropy calculations by these approaches seriously. These models are far from a stage in which they can provide a complete description of a black hole.

One can make somewhat more progress as regards the question of black hole evaporation though, again, the situation is not completely satisfactory. The role of black hole evaporation in the formation of black holes is also intimately related to the issue of information loss and it is probably best discussed in these contexts. We will now describe this issue which, as we said before, has led to some interesting controversies.

Consider a deck of cards arranged in a specific order. If you take this deck and throw it up against the wind, it will be scattered all over the floor. You might think the original information — viz., that the deck was ordered sequentially — is now completely lost and a person who sees the scattered cards can never retrieve it. This, however, is incorrect. If we were told how exactly the deck was thrown up, the wind pattern and other environmental details, one can, in principle, retrace the trajectory of each card lying on the ground. This is possible because the laws of mechanics are time reversible. When we do that, all the cards in the deck will go back and resume the initial position exactly as it was originally. A simple way of understanding this phenomenon is as follows. Imagine that a movie is made of your throwing the deck of cards into the wind, scattering them all over the floor. Micro reversibility essentially corresponds to running this movie backwards when you will see the cards on the ground flying back in the wind and forming the ordered deck. It is possible to do this mathematically and thus retrieve all the information about the initial state.

What if you throw the deck of cards into a black hole rather than throwing it up in the wind? How will you retrieve the information contained in the initial stage, once the deck has crossed the horizon? If you are willing to fall into the black hole, you may be able to retrieve some of it but then you still will not be able to convey it to the outside world. In fact, the situation is worse than that. If the matter is collapsing into a singularity, then the deck of cards is going to be crushed to a size less than Planck length just like any other matter and one does not even know what the concept of information is at these scales, let alone how to retrieve it. So, until we solve the problem of singularity in a viable quantum gravitational model, we cannot say anything about the final state of the deck of cards. It is fairly meaningless to talk about retrieving information about the initial state when we do not even know from what state this information has to be retrieved.

There is, however, a subtlety here. We have seen that, as far as the outside observer is concerned, the black hole takes an infinite time to form and the deck of cards just hovers outside the horizon. More to the point, all the matter and all the information which was contained in the initial state is distributed as a thin shell around the black hole horizon. If that is the case, the question of information loss may not have much to do with quantum gravity itself.

When we consider this closely, we realize that one cannot completely ignore the role played by the black hole evaporation. As the matter is collapsing to form a black hole, black hole evaporation is also proceeding as predicted by Hawking. This means that the total mass of the black hole is continuously decreasing, or rather, part of the mass which is falling into the black hole is reappearing at large distances in the form of particles produced by black hole evaporation. Obviously, if the matter which is falling into the black hole is taking part in the evaporation, then the information content in this matter should also be encoded in the radiated particles. There is, however, one serious difficulty with this idea. We saw that the particles which are produced and radiated to infinity have a thermal spectrum. It can be shown that a strictly thermal spectrum cannot encode any information; it is somewhat like a completely garbled alphabet or a random sequence of digits. Now we are in a bit of a problem. If there was no Hawking evaporation, we could have said that all the information in the initial state is stored in a final state which could even be a ball of matter with a radius equal to Planck length. But we know that at least part of the matter which forms the black hole contributes to the energy which is emitted as thermal radiation. It seems somewhat incredible that this will not entail any information loss. We once again see that the observer dependence of the phenomenon, existence of the singularity and the thermal nature of horizons conspire together to form a puzzle for which no satisfactory answer is available at present.

There is, however, another lesson which this analysis provides which is related to the idea we discussed in the last section. We said that Einstein's equations can be reinterpreted as a maximum entropy principle if we introduce some amount of gravitational entropy to the horizon. We did this by suggesting that, if the horizon is distorted, it could engulf some amount of normal matter and its entropy during such a distortion and to save the situation we need to assume that such distortions caused some amount of entropy. One could have thought of this result as attributing to every horizon certain degrees of freedom along with some entropy. In the case of a black hole, the concept of all matter and associated information hovering just outside the black hole reinforces the idea that some extra degrees of freedom need to be associated with the horizon.

Chapter 7

In the beginning ...

7.1 But, how did it all begin?

In this last chapter, we shall discuss what could possibly be considered *the* question human civilization has been asking ever since the beginning of recorded history — except that, as you will see, we now have lot more answers than we had possibly even fifty years back.

When we investigate the origins, we could worry about entirely different objects in nature differing widely in their past history. Since you might have wondered about many of these at one time or the other, it is probably worthwhile to begin by asking which particular cosmic object we are interested in tracking back in time. This is important because we do understand the origins of several cosmic objects very well — and even observe their formation directly — while others are still shrouded in mystery.

7.1.1 *Chemical elements*

The simplest of such questions will be related to the matter we see around us. We know that matter can be conveniently classified into ninety-odd chemical elements out of which more complex structures can be found. These chemical elements occur with different abundances in different cosmic sources but one finds that they have remarkable "integrity". That is to say, carbon found on Earth or found in an interstellar cloud behaves in an exactly similar manner as far as its physical and chemical properties are concerned. The origin of these elements, fortunately, is well understood.

We know that stars like the Sun shine because of nuclear reactions which take place in their interior. These reactions build heavier elements out of lighter ones by fusing the nuclei together. A sufficiently high mass star, during the late stages of its evolution, would have synthesized several heavy elements like carbon, nitrogen, oxygen etc. out of an initial composition made of essentially hydrogen and helium. At the end of its life cycle, such a high mass star explodes as a supernova spewing all the heavy elements into the interstellar medium. Planets like ours have formed out of such interstellar matter and, of course, contain heavier elements so essential for the origin and sustenance of life. Given the fact that nuclear fusion is an inherently quantum mechanical process, we once again note the quantum origin of the material world we see around us.

7.1.2 *Galaxies and stars*

We said that stars can synthesize heavy elements out of light ones and can lead to the familiar world around us. But how did the stars themselves form? For example, take the case of the Sun. Astronomical studies of different properties of the Sun using direct observations as well as theoretical modelling have allowed us to determine its age fairly precisely. The age of the Sun, from the time the nuclear reactions in it led to its shining, is roughly 4.5 billion years. A statement like this immediately leads to the question: How did the Sun form? Since the Sun is a typical star, this question can be rephrased in more general terms as: how do stars form?

Here again we are on rather firm ground. There are regions in our galaxy in which star formation is going on today, right before our eyes, so to speak. These occur in dense clouds of gas made of complex molecules. Every once in a while, a sub-region of a vast gas cloud contracts under its own weight, thereby becoming denser and denser. In a timescale of about a million years, under favourable circumstances, such a gas cloud will contract leading to central densities and temperatures which are adequate to trigger nuclear reactions. Once this happens, further gravitational contraction can be halted and a sort of equilibrium can be established. The nuclear reactions provide the energy which flows out of the gas cloud thereby making it a star. This is an ongoing process in our galaxy as well as in several external galaxies and hence can be directly observed through powerful telescopes. So, by and large, we understand how stars form.

Let us take a closer look at this process since it will also allow us to understand the formation of galaxies as well. After all, a galaxy is nothing but a collection of stars with either an elliptical distribution of stars or with stars mostly distributed in a plane. The former type of galaxies are called elliptical galaxies while the latter are called disc or spiral galaxies. What we need to understand is how such structures involving about a hundred billion stars can form out of a collapsing gas cloud with two broadly two distinct shapes — ellipticals and spirals.

To do this, we have to study the evolution of a system made of dark matter and normal matter bound by their own self-gravity. The dark matter, made of uncharged particles, will form a reasonably stable structure of definite size and mass. In the initial stages, the cloud of gas contained in it will be made of just hydrogen and primordial helium produced — in the big bang nucleosynthesis — with about 76 percent of weight being contributed by hydrogen. When this gas cloud contracts under the gravitational force of the dark matter, parts of gas will collide with each other and its temperature will increase making a fair amount of hydrogen to be ionized. Further evolution will depend on how efficiently the gas clouds can cool and form galaxy-like structures made of a collection of stars. We shall first describe a cooling process and fragmentation which could lead to the formation of stars; then we shall discuss how morphological differences like ellipticals and spirals can arise.

The free electrons and protons in the ionized gas can cool by two different processes. The first process is simple radiation of energy by the charged particles. Electrons and protons, while moving erratically in the gas cloud, can radiate photons thereby losing their energy. Secondly, the electrons can also lose their energy indirectly by colliding with neutral hydrogen atoms. When an electron collides with a hydrogen atom, it could kick the electron in the hydrogen atom from its ground state to a higher energy state, thereby transferring some of its energy to the hydrogen atom. The excited electron will, of course, fall back to the ground state within a short time, radiating away the extra energy. In the process some amount of kinetic energy which the original free electron possessed has been transfered to radiation.

These two processes help the gas cloud to cool down, but not by a large amount. It turns out that these processes act somewhat like a thermostat. If the gas cools too much, the free electrons will lose a lot of energy and will tend to get captured back by the proton thereby forming neutral atoms again. This process reduces the rate of cooling and tends to keep up the temperature. Similarly if the temperature increases too much, the collisional processes etc. also increase thereby increasing the rate of cooling; which, in turn, will cool the system more efficiently. Calculations show that these processes maintain the temperature at around 10,000 Kelvin.

Because the temperature does not rise, the pressure of the gas does not increase fast enough to prevent the gravitational forces from crushing the cloud. As a result, the gas cloud fragments into smaller pieces, so that within each small piece the pressure could balance gravity. (This works because the gravitational force increases with the size of the object; at a given temperature, the pressure can support a smaller object but not a larger one.) But this is a losing game. Soon the smaller fragments will go through the same process and the gravity will start dominating the thermal pressure in them. These fragments will soon break up into smaller pieces.

When can this avalanche stop? Remember that the loss of energy was possible only because the photons, emitted by the gas cloud, could escape from the system. If this does not happen and the photons are trapped inside, then the system cannot lose energy; pressure will build up and rapid fragmentation can be stopped. In other words, the gas cloud should become opaque (rather than transparent) to radiation. This will eventually happen when the density of the gas cloud is sufficiently high.

Once the pressure forces are comparable to self-gravity, the fragmentation stops and each piece enters a stage of very slow contraction. They radiate at a very reduced rate because of the increased opaqueness to photons. Such an object is called a "protostar" and grows in mass as more matter is accreted from the surroundings. A protostar will, in general, be also spinning because of the residual gas motion. The matter which accretes to the protostar will form a flattened, spinning, "protostellar disc" of gas around it. The slow gravitational contraction of the protostar will continue until the temperature is high enough for nuclear reactions to take place at the centre. And then, a star will be born! The dense gaseous disc around a protostar will undergo its own dynamics and there is every likelihood that planetary systems orbiting the star can form out of it.

The above process could account for the formation of a collection of stars starting from a gaseous cloud. Depending on the distribution of these stars, such a collection could look like an elliptical or a spiral. This broad division of galaxies arises as follows. When a rotating gas cloud contracts under its own weight, it will have a tendency to get flattened in the direction perpendicular to the axis of rotation. If the rate of star formation is slow compared to the time taken for the gas to collapse, most of the gas will end up in a disc perpendicular to the direction of rotation *before* star formation takes place. In such a case, most of the stars will form in the disc and we will end up with a disc galaxy. On the other hand, if the star formation rate is high, stars will form even before the gas can collapse to a disc. Once star formation has progressed significantly each star will move in its own orbit in the common gravitational field of the system and further collapse will not take place. In this case, we will end up with a more spread out distribution of stars like in an elliptical.

Another equivalent way of understanding this difference is the following. When a star gets formed, it retains the speed with which it is born and moves around in the gravitational field of other stars and gas in some specific orbit. If the star was formed before the gas collapsed to a disc, it will move in some random direction. On the other hand, when the gas collapses to form a disc, it will lose most of its speed perpendicular to the disc and any star which gets formed *afterwards* will only have motion in the plane of the disc. Thus elliptical galaxies have stars moving in a fairly spread-out fashion while disc galaxies have stars moving mostly in the plane.

We, therefore, conclude that the elliptical galaxies formed their stars very efficiently in a dramatic burst during the initial collapse of the gas cloud while the spiral galaxies probably formed only ten percent of their stars in the initial phase. Most of the stars in a spiral galaxy formed in a leisurely fashion spread over the past ten billion years after the gas had already formed the disc. As a consequence, most of the light in an elliptical is from older stars while the spirals shine mostly due to younger stars.

How can we understand the formation of larger structures like groups and clusters? Here we need to take into account yet another factor: mergers of galaxies. In today's universe, collisions and mergers between galaxies are rare events but not so in the past. Since structures developed hierarchically, we would have expected single galaxies to form before large groups or clusters. Computer simulations of such systems have shown that the merging of several smaller sub-units can lead to the formation of nearly uniform looking spherical clusters. Similarly, clouds with masses of 10^6 to 10^7 $M_\odot$ used to merge frequently when the universe was about a tenth of its present age and formed individual massive galaxies. Since such merging cannot be 100 percent efficient, one would expect to find the massive galaxies to be surrounded by several dwarf-like companions. This is precisely what we see in the local group, which is dominated by Milky Way and Andromeda, each of which is surrounded by several dwarf galaxies.

7.2   The origin of initial perturbation

The picture of galaxy formation described above rests on one crucial input. It assumes that very early in the evolution of the universe there existed some fluctuations or unevenness in the distribution of matter. Some regions had slightly more matter and some regions had slightly less. Given this initial condition, gravity can do the rest of the job. Regions which are overdense will exert greater gravitational pull and will grow denser while the regions which had less density will lose out in the process. Eventually, one will end up with the kind of uneven distribution of matter concentrated in the form of galaxies and clusters of galaxies which we see today.

You must remember that there is very strong direct evidence for the above idea. As we described in Chapter 5, we receive radiation from an epoch when the universe was a thousand times smaller in size. In other words, a fossilized photograph of the state of the universe at the time when it was a thousand times smaller has been made available to the astronomers by Nature! By studying this photograph, one can directly verify that there were small fluctuations in the density at that time. Given this initial condition — at the time when the universe was a thousand times smaller — one can evolve the dynamics forward in time in a clear cut manner. In fact, cosmologists routinely simulate the growth of structure in the universe using powerful computers and these studies show that we can indeed form the kind of universe we see around us starting from a time when the universe was 1000 times smaller.

This is nice but we next need to know why there were small fluctuations when the universe was 1000 times smaller. Here, unfortunately, we do not have direct evidence but one plausible model that leads to such fluctuations is based on the idea of an inflationary universe. As we outlined in Chapter 5, the inflationary model postulates that the universe underwent a very rapid phase of expansion very early in the universe when it was barely 10^{-34} seconds old. At that time, the description of matter in the universe has to be quantum mechanical. We have already seen in Chapter 4 that the vacuum in quantum theory can be perturbed by external conditions. The rapid expansion of the universe has the effect of ripping apart the vacuum of the quantum fields and producing virtual particles. This is very similar to the processes like the particle production by an electric field or a black hole which we discussed in Chapter 6. The resulting fluctuations in the energy density of matter act as the seed for further growth.

This idea suggests, once again, that *our origins are purely quantum mechanical*. It is the intrinsic quantum nature of the vacuum state which leads to the fluctuations in the energy density during the inflationary phase. The rapid expansion of the universe during the inflation has the effect of — loosely speaking — amplifying these quantum fluctuations to fluctuations which can be described by classical gravitational physics. They grow further due to the gravitational instability we described earlier and finally lead to the tell-tale signs in the cosmic microwave background radiation.

7.3 Quantum cosmology

There is a well known story of an elderly lady approaching a cosmologist and claiming that the Earth is supported in space by a giant turtle. The cosmologist tried to argue against it by pointing out that the turtle will need a support in her model. The lady replies that "it is supported by another, bigger turtle". The exasperated cosmologists enquires how the second turtle is supported and gets the reply "it is turtles all the way"! You might think that the scientific explanation for the origin of the universe is also by and large "turtles all the way" — except that the turtle is essentially quantum theory.

Nowhere does this idea strike you with greater force than when one starts asking questions about the origin of the universe itself. It is all fine *to start* with an expanding universe, maybe with an inflationary phase thrown in, generate perturbations in the distribution of matter density from inherent quantum fluctuations, make them grow via gravitational instability, form the galaxies and stars, generate heavier elements inside the stars and throw them around in supernova explosions, and finally lead to the formation of planetary system and life. But how did we get the universe going in the first place?

This is, of course, the riddle of creation that has plagued humanity since the beginning of its enquiry. To be sure, we do not yet have the answers with any certainty but we do have a paradigm in which such questions can be formulated scientifically and analyzed mathematically. You have to realize that this is significant progress by itself even though, as I said above, we do not yet have The Final Answer. Is physics on its way to explain the process of creation, possibly the ultimate riddle for humanity? The answer, of course, depends on what you mean by "explain" and it requires somewhat detailed clarification.

In the development of any physical theory, one has to distinguish between different phases. The first phase involves formulating the problem in precise mathematical terms so that one can be quite sure what the discussion is all about. The second phase is to provide a theoretical framework in which one can hope to address the problem which has been presented. The last phase, of course, is the actual solution of the problem by its mathematical treatment and its verification by comparison with observations.

Take a simple problem like the turbulent flow of water at the bottom of a waterfall. If we want to study the actual pattern of flow there, you have to go through all the three different phases mentioned above. One starts by defining mathematical objects like a vector at each point in space which denotes the velocity of the water etc. One then needs to know the mathematical equations — in this case the equations of fluid mechanics — which will govern the behaviour of various mathematical quantities that we have introduced. Finally, one needs to solve these equations and determine how exactly the water from the waterfall flows.

It is quite appropriate to assume that, in this particular example, there are no real unresolved issues as regards the first two phases. We actually know what the variables which need to be determined are and we are also quite confident about the equations which govern their behaviour. It turns out that the third phase of this problem is impossibly difficult even for the largest computers available today. This is just a limitation in our ability to solve complicated mathematical equations. It is therefore conventional to introduce various approximate models in which the relevant equations can be solved either analytically or in a computer. This is in fact what an aeronautical engineer, faced with the issue of studying the turbulent motion of air around the aircraft wing, will do.

Move on from the waterfall to the creation of the universe. The situation here is different as regards all the three phases, but at varying degrees. To begin with, we are broadly clear as to what the problem is. We do have a description for the universe starting from a very tiny fraction of a second all the way up to several billions of years. What we need is a description which will tie up with this at the earliest phase. The problem here is again that of bringing together the principles of quantum theory and gravity.

Given the equations in physics determining the behaviour of a system — whether it is a ball thrown from the Earth or the universe itself — one can determine both the past and future evolution from a given moment of time. For example, if you know the position and speed of a ball which was thrown from the surface of the Earth at a given moment of time, you can not only predict what the ball will do in future but also what it did in the past. Similarly, given the parameters of the present day universe and the equations governing its behaviour, you can figure out not only how the universe will behave in future but also how it got to the present stage. When we try to do that in a purely classical description of gravity, based on Einstein's theory, one encounters a nonsensical behaviour for the mathematical equations at the earliest moment. Any acceptable cosmological model predicts that the universe had infinite density and temperature at some finite time in the past. Such an infinity — called a mathematical singularity — prevents us from determining what happened before that event which is usually christened as "big bang". This is understandable, though disheartening, because we do expect the classical Einstein's equations to break down when the temperatures are very high. Once again, we need a quantum mechanical description of gravity which we do not have. So, even though we broadly know the questions that we need to address, we do not have a mathematical formalism in which these questions can be rigorously framed.

One could, of course, just sit around and wait for somebody to develop a complete quantum theory of gravity and hope that the issues related to creation will get solved in such a framework. There are several good reasons not to do this, though. For one thing, physics progresses by making bold extrapolations and educated guesses even before we have a complete theory for a phenomenon. It is also not clear that even if we have a complete theory of quantum gravity, we can immediately figure out the answers to all the questions. As we said before, this has not been possible even for fluid mechanics where we do know the relevant laws of physics. So, some kind of approximate model building will have a value even if we have the complete theory for quantum gravity.

The most important reason why we may want to do this, however, is to develop the framework based on which the relevant questions can be asked. This is important because one might be overawed by the task of explaining everything about the universe and might wonder whether this is really in the realm of science. It is, therefore, gratifying that people could develop mathematical frameworks, loosely called quantum cosmological models, in which one could pose the relevant questions and try to answer them. These models are just that, viz., models. What they do is to take the standard description in Einstein's theory for a cosmological model and try to put it in a quantum mechanical language. For example, the size of the universe and the rate at which the size is increasing will be modelled somewhat like the position and speed of a particle. One then tries to translate what we know in normal quantum mechanics regarding the description of a particle to understand the quantum behaviour of the universe.

Obviously, you cannot demand too much of such models. In the ultimate analysis, a quantum cosmological model should tell you all the parameters of the universe and we are nowhere close to determining this. In fact, we do not even know how to phrase such a question. The success of quantum cosmological models, however, is in a different domain. First, they showed that it is indeed possible to circumvent the singularity in the classical theory by introducing quantum mechanical concepts. Roughly speaking, the singularity encountered in classical theory corresponds to a universe with zero size and zero rate of expansion. But since the size of the universe and the rate of expansion are analogous to the position and speed of a particle, we know from the uncertainty principle that they cannot be determined simultaneously with arbitrary accuracy in any quantum description. So you really cannot have the classical singularity in a purely quantum mechanical context. While this appears reasonable, the models are not powerful enough to tell you what exactly replaces the singular behaviour of the classical universe. The second, and probably more interesting question addressed in some of these models, has to do with the creation of matter and spacetime. Since this fascinates many people, let me describe how it comes about in these models.

7.4　Cosmogenesis

To see what is involved, let us probe more closely the conventional difficulty in "creating the universe". By and large, one would have said that such an idea violates our concepts of conservation of certain physical quantities. In every day life, we never see television sets or tea cups getting created and popping out of nothing; so how is it that all the matter in the universe could have come into being in a manner which does not violate any of the cherished notions of physics?

If you have been following the discussion above, you would have realized that we really do not have to produce television sets, tea cups or even stars or galaxies "out of nothing". We have a reasonably well defined paradigm which can start from the quantum phase of the initial universe and generate all these. What is more, such a paradigm has sound observational support. What we really need to understand is how one could create spacetime, and the quantum fields which live on the spacetime, at a sufficiently early epoch.

Normally, one would have thought that this is banned by the existence of certain conservation laws in physics. For example, the total electric charge or the total angular momentum or energy in the universe is conserved and we can never generate them without violating the known laws of physics. On closer inspection, however, we notice something truly remarkable. Observations show that all the fundamental conserved quantities in physics seem to have vanishing values for our observed universe. To understand the significance of this statement, let us consider the idea of angular momentum. Observations show that our universe at the largest scales does not rotate even though virtually every individual object in the universe — planets, stars, galaxies, ... — has an intrinsic rotation. There are cosmological solutions to Einstein's equations which describe a universe which is rotating on the whole. But the real universe we see does not belong to this class of solutions and instead follows closely the one which does not rotate. If the universe had a sense of rotation, it would posses nonzero angular momentum. Starting from a vacuum state which has zero angular momentum one cannot reach a state with nonzero angular momentum without violating a conservation law in physics. This problem does not arise because the universe so conveniently has zero angular momentum.

The same is true as regards electric charge, for example. All the laboratory experiments show that one cannot create or destroy net electric charge. Nothing prevents the existence of the universe which has some amount of net electric charge. If our universe had net electric charge, there is no way we could produce it 'out of nothing'. But — again — observations show that our universe has zero net electric charge just like the vacuum state.

The real crunch comes when we study the total amount of matter which, according to relativity, is equivalent to the total amount of energy. Creating the matter out of the vacuum is, of course, possible under suitable conditions. Given the fact that galaxies in the universe are moving away from each other with a significant amount of speed, one might have thought that there is also a huge amount of kinetic energy associated with the expansion of the universe. Any model for cosmogenesis, one would have said, needs to account for the origin of this vast amount of energy. Once again, it turns out that the net energy of the universe is very close to zero! This comes about because the gravitational attraction among the constituents of the universe contributes a *negative* gravitational potential energy which precisely compensates for the positive kinetic energy of the expansion. Taken together, the universe has zero energy just like the vacuum state. (I should add that this argument is somewhat oversimplified since neither the kinetic energy of expansion nor the total gravitational potential energy can be rigorously defined in general relativity for an expanding universe. But the qualitative argument captures the essence of the situation though it is incorrect technically.)

Given that all the conserved quantities for the observed universe have the same value as for the canonical vacuum state, one is lead to an intriguing possibility that the material universe could actually emerge out of "nothing"; that is, out of a vacuum state as a quantum fluctuation. Such a model for cosmogenesis works roughly as follows:

We consider, in a purely mathematical manner, two different quantum states. The first one is a state in which the spacetime is flat and all quantum fields reside in their vacuum states; the second one has an expanding universe with a certain amount of matter populating it. We then consider the possibility that, treated as a quantum mechanical system, a transition occurs from the first state to the second. A first step in such an analysis is to ensure that such a transition does not violate any known conservation law which would have immediately forbidden the occurrence of such a transition. This is what we argued for in the discussion above.

The next question — which is more complex to answer — is what triggers such a transition. Here the key idea is again that of gravitational instability of the vacuum state itself. Consider a region of spacetime in which spontaneous quantum fluctuations lead to the localization of some amount of energy. This will lead to a corresponding fluctuation in the gravitational field. We now have a quantum field coupled to an external gravitational field which, in principle, is capable of producing particles out of vacuum. Once this happens, the process can be self sustaining and an instability can be triggered. In this picture we consider the second state in which there is a nonzero amount of matter and an expanding spacetime to be the result of instability of the initial state. This instability creates matter and in some sense the spacetime itself.

It is gratifying, though at times a bit perplexing, that creation itself is a variation on the quantum theme just as several other phenomena which have all played a crucial role in making our universe what it is.

Index